Oxford Applied Mathematics and Computing Science Series

General Editors
J. Crank, H. G. Martin, D. M. Melluish

BERNARD CARRÉ
University of Southampton

Graphs and networks

CLARENDON PRESS · OXFORD
1979

Oxford University Press, Walton Street, Oxford OX2 6DP

OXFORD LONDON GLASGOW NEW YORK
TORONTO MELBOURNE WELLINGTON IBADAN
NAIROBI DAR ES SALAAM LUSAKA CAPE TOWN KUALA LUMPUR
SINGAPORE JAKARTA HONG KONG TOKYO DELHI
BOMBAY CALCUTTA MADRAS KARACHI

© Oxford University Press 1979

British Library Cataloguing in Publication Data

Carre, Bernard
   Graphs and networks.—(Oxford applied
mathematics and computing science series).
1. Graph theory
I. Title   II. Series
511'.5   QA166   79–40423

ISBN 0 19 859615 4
ISBN 0 19 859622 7 Pbk

QA
166
C38

PRINTED IN GREAT BRITAIN BY
J. W. ARROWSMITH LTD., BRISTOL, ENGLAND

To
Françoise, Isabelle, and Claire

Preface

'DISCRETE SYSTEMS' or organized collections of objects are frequently encountered, for instance in computer science, engineering, and industrial management. Graph theory and network flow theory provide simple techniques for constructing models of systems of this kind, and powerful methods for their analysis and optimization.

The object of this book is to provide a simple account of the basic results and techniques of graph and network flow theory, with strong emphasis on their applications and algorithmic aspects.

The book is written for undergraduate or first-year graduate students in mathematics, computer science, engineering, or operational research. Students with some background in 'modern' algebra should find the book very easy to read; for those who have no previous experience of modern mathematics, the first chapter is intended to provide a simple but adequate introduction to algebraic structures. There are no other prerequisites, although some knowledge of linear algebra—in particular of the basic direct and iterative methods of solving systems of linear equations—would be useful in reading the chapter on path problems (and, I hope, might make it a little amusing).

Although the presentation is rather informal, all the results given in the main text are proved, and a justification is given for all but the most trivial algorithms.

Practical applications in computer science, engineering, and operational research are presented, for the most part, as examples in the body of the text. To make these realistic it has sometimes been necessary to enter into technical details, but it is nevertheless hoped that all the examples will be easily comprehensible, to readers of any specialization.

With regard to the methods of presentation of the graph theory, the most novel feature of the book is the algebraic approach to path problems and connectivity, in Chapters 3 and 4. Path problems arise in many different forms, and the variety of algorithms invented to solve them is at first sight quite bewildering. The algebraic approach to this subject provides a simple, elegant framework in which to

situate a host of path problems, and allows a systematic development of general algorithms to solve them.

In teaching this material I have found that mathematics students
are pleased—and sometimes even a little surprised—to discover
that abstract algebra can have mathematically substantial and at the
same time very practical applications. Engineering students are less
used to abstract structures but they nevertheless find the setting
familiar: the matrices which we use to describe graphs have the
same 'structure' as the admittance matrices of electric circuits and
the stiffness matrices of civil-engineering frameworks, and our
methods of solving linear systems are analogous.

In this context it will be observed that, although some authors
classify path-finding algorithms according to whether they are
'algebraic' or 'matrix' methods or 'other' methods, I do not accept
this type of classification. Algebras are *languages* in which we
formulate problems and devise methods for their solution. The
ultimate usefulness of an algebraic approach depends on the ease
and elegance with which it enables us to formulate problems
precisely, and to derive all the 'best', i.e. most efficient methods of
solving them. Here we give an algebraic development of all the most
important path-finding methods, and we even unashamedly present
critical path analysis as a problem of solving a system of simultaneous equations. Whether or not the algebraic approach meets its
objective will be for the reader to judge. Of course, to implement a
method on a computer, we must represent the mathematical objects
which it involves by appropriate data structures (for instance, we
may represent a sparse matrix by linked lists), and a computer's
view of a method will always be much the same, whichever formal-
ism we use to develop it.

With regard to our particular choice of algebraic structure for
discussing path problems, it was necessary to make a com-
promise, between achieving maximum generality on the one hand
and clarity of exposition on the other. The directions in which
further generalizations can be made are indicated at the close
of Chapter 3.

To present network flow methods, in Chapter 6, I have made
extensive use of the notion of a 'displacement network', which
defines the 'perturbations' which can be made to a given network
flow. This approach, quite often used in operational research, seems
particularly appropriate for teaching the subject to engineers: the

displacement network plays essentially the same role as the 'small-signal equivalent circuit' which is familiar to every student of electronics.

The choice of a language in which to present algorithms is not easy. The development of an algorithm for a computer involves four stages, (i) the development of a mathematical model of the problem to be solved, (ii) the formulation of a solution method in terms of the mathematical objects defined in the first stage, (iii) the choice of data structures to represent the mathematical objects in the computer memory (for instance, we may represent a set by a binary vector, a linear list, or a stack), and (iv) programming the method, in terms of the chosen data structures.

Since this book is mainly concerned with aspects (i) and (ii), algorithms are mostly described in plain English and familiar mathematical notation. However, since the choice of data structures can strongly affect the efficiency of graph algorithms, this matter is taken up in a few instances. Where there is a choice of solution methods, I have tried to explain the principles of those which can be implemented most efficiently; where possible, I have also given references to program descriptions.

'Backtrack programming' or tree-search algorithms can be described most concisely and elegantly in a recursive form, but I felt this would not be sufficiently transparent to most readers; however, I have tried to describe each tree-search algorithm in such a way that, for readers familiar with recursive programming, the outline of the recursive procedure would be clearly visible.

Since 1970, computer scientists have made important contributions to graph theory by developing the notion of computational complexity; they have also achieved remarkable improvements in performance of graph algorithms, mainly through the clever manipulation of data structures. The literature on some of these refined algorithms makes remarkably little use of graph theory: the validity proofs are written in the language of computer science (procedure calls, depth of recursion, popping and pushing nodes onto a stack). I have not attempted to give full details of these procedures; for these the reader will be referred to the original papers. However, the ways in which their authors use data structures sometimes have graph-theoretic interpretations, which may aid comprehension; where I have been able to make such interpretations, I have given them.

x *Preface*

In writing this book I have benefited from the help of many people. In particular, I have greatly enjoyed and profited from collaborative work on path algebras with Roland Backhouse and Ahnont Wongseelashote, and their ideas strongly influenced me in writing Chapter 3. They also made many valuable comments on other parts of the manuscript. Jean-François Bergeretti checked the entire manuscript, and I am greatly indebted to him for all his advice and help. I also received useful comments from Keith Lloyd, and from several of my M.Sc. students, especially David Gill and Steven Scott.

The manuscript was beautifully typed by Gina Pugsley and Elaine Hare. I would also like to thank the staff of the Oxford University Press for their encouragement and forbearance.

Finally, I am grateful to Françoise, my wife, not only for her proof-reading but also for years of patience, and latterly some impatience, without which this book would never have been completed.

Southampton, July 1978   B.C.

Contents

1 Algebraic foundations

1.1. Sets

1.1.1. The notion of set

A *set* is a collection of distinct objects of any nature, which are called its *elements* or *members*. The following are examples:

> The set of all fish in the Monaco Aquarium.
> The set of all International Phonetic Script symbols.
> The set of Seven Deadly Sins.

These are all *finite sets*, i.e. sets having only a finite number of elements. As an example of an infinite set we have:

> The set of all positive integers.

Here it will usually be convenient to represent sets by capital letters $A, B, C, \ldots$, and elements of sets by lower case letters $a, b, c, \ldots$. To indicate that an object x is an element of a set A we use the notation

$$x \in A,$$

which is read as 'x belongs to A'. If x is *not* an element of A, we may write

$$x \notin A,$$

which is read as 'x does not belong to A'.

A set is completely determined by its elements, i.e. it is fully described by specifying which objects belong to it. Two sets A and B are said to be *equal*, $A = B$, if they have the same elements, i.e. if every element of A is an element of B and every element of B is an element of A.

The number of elements in a finite set A is called the *cardinality* of A and is denoted by $|A|$.

1.1.2. The specification of sets

Particular sets are usually specified in one of two ways. The first method is to list all the elements of the set, between braces. For example, to specify a set S whose elements are the integers 2, 3, 5, and 7 we may write

$$S = \{2, 3, 5, 7\}.$$

The order in which the elements are listed has no significance. Thus $\{2, 3, 5, 7\} = \{3, 7, 5, 2\}$.

Alternatively, it is often convenient to specify a set in terms of a property which is *characteristic* of its elements, i.e. a property which all elements of the set and only elements of the set possess. In this case the following type of notation is used:

$$P = \{x \mid x \text{ is a prime number}\}.$$

This is read as 'P is the set of all objects x such that x is a prime number', or more simply as 'P is the set of all prime numbers'. In this notation the symbol on the left of the vertical line $\mid$ stands for a typical element of the set, while on the right of the vertical line is a statement about this typical element which serves to determine the set.

Example 1.1. The set $\{q \mid q \text{ was a wife of Henry VIII}\}$ has the following elements: Catherine of Aragon, Anne Boleyn, Jane Seymour, Anne of Clèves, Catherine Howard, Catherine Parr.

Example 1.2. Let $P = \{x \mid x \text{ is a prime number}\}$. Then the set $S = \{2, 3, 5, 7\}$ can be defined by the statement

$$S = \{x \mid x \in P \text{ and } x < 8\},$$

or more concisely, by the statement

$$S = \{x \in P \mid x < 8\},$$

which is read as 'S is the set of all elements x of P such that x is less than 8'.

1.1.3. Subsets

Given two sets A and B, the set B is said to be *subset* of A if every element of B is an element of A. The statement 'B is a subset of A' is written symbolically as

$$B \subseteq A \quad \text{or} \quad A \supseteq B.$$

It will be noted that according to the above definition, a set A is always a subset of itself, $A \subseteq A$. Any subset B of A which is *not* equal to A is called a *proper* subset of A. The statement 'B is a proper subset of A' is denoted by

$$B \subset A \quad \text{or} \quad A \supset B.$$

If $A \subseteq B$ and $B \subseteq A$, then every element of A is an element of B and every element of B is an element of A. Hence, from our definition of equality,

$$\text{if} \quad A \subseteq B \quad \text{and} \quad B \subseteq A \quad \text{then} \quad A = B.$$

In later sections we shall sometimes be faced with the problem of proving that two sets are equal; the above statement suggests a two-pronged attack on such a problem, for it follows from this statement that to prove the equality of two sets A and B, it is sufficient to demonstrate that (i) $A \subseteq B$ and (ii) $B \subseteq A$.

1.1.4. Union and intersection

There are various ways of combining sets to form other sets, which will be presented in this and the following sections.

If A and B are sets, then their *union* $A \cup B$ is defined by

$$A \cup B = \{x \mid x \in A \text{ or } x \in B\}$$

and their *intersection* $A \cap B$ is defined by

$$A \cap B = \{x \mid x \in A \text{ and } x \in B\}.$$

In other words, the union $A \cup B$ is the set of all elements which belong to A or to B (or to both); whereas the intersection $A \cap B$ is the set of all those elements common to both A and B.

Example 1.3. If $S = \{2, 3, 5, 7\}$ and $T = \{1, 2, 3\}$, then $S \cup T = \{1, 2, 3, 5, 7\}$ and $S \cap T = \{2, 3\}$.

If two sets A and B have no element in common, they are said to be *disjoint*.

It will be noted that if A and B are disjoint then the set $A \cap B$ does not contain any elements. However, for our purposes the concept of a set which contains no elements is quite acceptable, and indeed very useful: we shall call this set the *empty set* and denote it by the symbol ϕ. Thus if A and B are disjoint, $A \cap B = \phi$. The empty set is a subset of any set A. (Indeed, since ϕ does not contain any elements at all, it does not contain any elements which do not belong to A.)

The operations of union and intersection have several important properties. First, it is evident from the definitions of these operations that they obey the *idempotent laws*

$$A \cup A = A \quad \text{and} \quad A \cap A = A,$$

and the *commutative laws*

$$A \cup B = B \cup A \quad \text{and} \quad A \cap B = B \cap A,$$

for all sets A, B.

They also obey the *associative laws*

$$(A \cup B) \cup C = A \cup (B \cup C) \quad \text{and} \quad (A \cap B) \cap C = A \cap (B \cap C),$$

and the *distributive laws*

$$A \cup (B \cap C) = (A \cup B) \cap (A \cup C)$$

and

$$A \cap (B \cup C) = (A \cap B) \cup (A \cap C),$$

for all sets A, B, and C. These identities can be proved formally by the method suggested in Section 1.1.3, which is demonstrated in the next example.

Example 1.4. To prove, that $A \cap (B \cup C) = (A \cap B) \cup (A \cap C)$, for all sets A, B, C.

Proof. (i) If $x \in A \cap (B \cup C)$, then $x \in A$, and $x \in B$ or $x \in C$. In the first case $x \in A$ and $x \in B$, hence $x \in A \cap B$. In the second case $x \in A$ and $x \in C$, hence $x \in A \cap C$. Hence in both cases, $x \in A \cap B$ or $x \in A \cap C$, so that $x \in (A \cap B) \cup (A \cap C)$. This proves that

$$A \cap (B \cup C) \subseteq (A \cap B) \cup (A \cap C).$$

(ii) Conversely, if $x \in (A \cap B) \cup (A \cap C)$ then $x \in A \cap B$ or $x \in A \cap C$. In the first case $x \in A$ and $x \in B$, and similarly, in the second case $x \in A$ and $x \in C$. Hence in both cases $x \in A$ and $x \in B \cup C$, so $x \in A \cap (B \cup C)$. This proves that

$$(A \cap B) \cup (A \cap C) \subseteq A \cap (B \cup C).$$

Combining the final relations of paragraphs (i) and (ii) we conclude that $A \cap (B \cup C) = (A \cap B) \cup (A \cap C)$.

1.1.5. Difference and complement

If A and B are sets then the *difference* $A - B$ is defined by

$$A - B = \{x \mid x \in A \text{ and } x \notin B\}.$$

In other words, the difference $A - B$ is the set of all those elements of A which do not belong to B. If B is a subset of A then $A - B$ is sometimes called the *complement of B in A*.

Example 1.5. If $S = \{2, 3, 5, 7\}$ and $T = \{1, 2, 3\}$ then $S - T = \{5, 7\}$ and $T - S = \{1\}$.

1.1.6. Ordered pairs, and products of sets

In a game of tennis the score is declared simply by saying, for instance, 'forty, thirty'. Such a declaration is unambiguous because it is understood that the first number of the pair is the score of the person serving the ball, and the second the score of his opponent. Thus if one declared the score as 'thirty, forty' when one should say 'forty, thirty', the person serving the ball would probably complain. In mathematical terms, the score in a game of tennis is an example of an *ordered pair* of objects (which in this case are numbers).

If x and y are objects, we shall denote by (x, y) the pair consisting of x and y in that order. Two ordered pairs (u, v) and (x, y) are said to be *equal* if and only if $u = x$ and $v = y$. We describe x as the *first component* and y as the *second component* of (x, y).

If A and B are sets, then the *Cartesian product* $A \times B$ of A and B is the set of all ordered pairs (x, y) such that $x \in A$ and $y \in B$. In symbols,

$$A \times B = \{(x, y) \mid x \in A \text{ and } y \in B\}.$$

Example 1.6. Let $P = \{a, b\}$ and $Q = \{1, 2, 3\}$. Then

$$P \times Q = \{(a, 1), (b, 1), (a, 2), (b, 2), (a, 3), (b, 3)\};$$
$$Q \times P = \{(1, a), (1, b), (2, a), (2, b), (3, a), (3, b)\};$$
$$P \times P = \{(a, a), (a, b), (b, a), (b, b)\}.$$

1.1.7. Sets of sets

The notion of a set whose elements are themselves sets will not be entirely unfamiliar: For instance, the European Economic

Community is a set of sets (nations) of people. In later sections we shall frequently encounter sets of sets, and in particular sets of the kind defined below:

If A is any set, then the *power set* $\mathcal{P}(A)$ of A is the set of all subsets of A. Thus

$$\mathcal{P}(A) = \{X \mid X \subseteq A\}.$$

Example 1.7. Let $A = \{x, y, z\}$. Then

$$\mathcal{P}(A) = \{\phi, \{x\}, \{y\}, \{z\}, \{x, y\}, \{x, z\}, \{y, z\}, \{x, y, z\}\}.$$

The operations of union and intersection can be defined for a set of sets, as follows:

Let $\mathcal{S}$ be any set of sets. Then the *union* $\cup \mathcal{S}$ is the set of all objects which belong to at least one member of $\mathcal{S}$, that is

$$\cup \mathcal{S} = \{x \mid x \in A \text{ for at least one } A \in \mathcal{S}\},$$

and the *intersection* $\cap \mathcal{S}$ is the set of all objects which belong to all the members of $\mathcal{S}$, that is

$$\cap \mathcal{S} = \{x \mid x \in A \text{ for all } A \in \mathcal{S}\}.$$

If $\mathcal{S}$ is a finite set of sets, say $\mathcal{S} = \{A_1, A_2, \ldots, A_n\}$, then in place of the notation $\cup \mathcal{S}$ we often write

$$\bigcup_{k=1}^{n} A_k \quad \text{or} \quad A_1 \cup A_2 \cup \cdots \cup A_n$$

and similarly, in place of $\cap \mathcal{S}$ we write

$$\bigcap_{k=1}^{n} A_k \quad \text{or} \quad A_1 \cap A_2 \cap \cdots \cap A_n.$$

Example 1.8. Let $\mathcal{S} = \{A, B, C\}$ where $A = \{2, 3, 5, 7\}$, $B = \{1, 3, 5\}$, and $C = \{1, 2, 3\}$. Then $\cup \mathcal{S} = \{1, 2, 3, 5, 7\}$ and $\cap \mathcal{S} = \{3\}$.

As we have seen, the power set $\mathcal{P}(A)$ of a set A is constructed by collecting together *all* the subsets of A. However, there are other ways of constructing new sets from subsets of A, the following being of special importance:

A set $\mathcal{S}$ of non-empty subsets of a set A is called a *covering* of A if each element of A belongs to at least one member of $\mathcal{S}$, i.e. if

$$\cup \mathcal{S} = A.$$

A covering $\mathscr{S}$ of A is described as a *partition* of A if it has the additional property that all distinct pairs of elements of $\mathscr{S}$ are disjoint. In other words, $\mathscr{S}$ is a partition of A if each element of A belongs to one and only one member of $\mathscr{S}$, and thus $\mathscr{S}$ 'decomposes' A into various 'parts'.

Example 1.9. If $S = \{a, b, c, d, e\}$, then $\{\{a, b\}, \{b, c, d\}, \{b, c, e\}\}$ and $\{\{a, b\}, \{c, d, e\}\}$ and $\{\{a\}, \{b\}, \{c\}, \{d\}, \{e\}\}$ are three coverings of S, the last two being partitions.

Example 1.10. Let A be the set of letters of the alphabet:

$$A = \{a, b, c, \ldots, z\}.$$

We may 'classify' these letters, according to whether they are vowels or consonants, obtaining the two 'classes' of letters

$$V = \{x \in A \mid x \text{ is a vowel}\} = \{a, e, i, o, u\},$$

$$C = \{x \in A \mid x \text{ is a consonant}\} = \{b, c, d, f, \ldots, z\}.$$

The set $\{V, C\}$ is a partition of A. Any 'classification' of a set of objects such that each object falls into one and only class results in a partition.

Exercises

1.1. Prove that for any sets A and B, the three conditions $A \subseteq B$, $A \cup B = B$, and $A \cap B = A$ are mutually equivalent.

1.2. Prove that for any sets A and B,
$$A \cup (A \cap B) = A \quad \text{and} \quad A \cap (A \cup B) = A.$$

1.3. Prove that
$$A - (B \cup C) = (A - B) \cap (A - C)$$
and
$$A - (B \cap C) = (A - B) \cup (A - C).$$

1.4. For any sets A and B, we define the *symmetric difference* of A and B as the set
$$A \triangle B = (A \cup B) - (A \cap B).$$

Prove the following identities:

  (i) $A \triangle B = B \triangle A$
  (ii) $A \triangle A = \phi$
 (iii) $(A \triangle B) \triangle C = A \triangle (B \triangle C)$
 (iv) $A \cap (B \triangle C) = (A \cap B) \triangle (A \cap C).$

1.2. Functions

1.2.1. The notion of function

Let A and B be any sets. Then a *function f from A to B* is determined by any rule which assigns to each element x of A a single element of B; this element of B is called the *image of x under f*, and is denoted by $f(x)$.

It will be convenient to use the notation

$$f:A \to B$$

to indicate that f is a function from A to B. The set A is called the *domain* of f, and B its *codomain*. The set $f(A)$ of all images $f(x)$ of elements $x \in A$ is called the *range* of f. Note that $f(A)$ may be a proper subset of B.

Example 1.11. Let P be the set of all living people and let $\mathbb{N}$ be the set of non-negative integers. Then we can define a function f from P to $\mathbb{N}$ by the rule: *if $x \in P$ then $f(x)$ is the age in years of x.*

Example 1.12. Let $\mathbb{R}$ be the set of real numbers, and $\mathbb{R}^+$ the set of positive real numbers. Then we can define a function from $\mathbb{R}$ to $\mathbb{R}^+$ by the rule: *if $x \in \mathbb{R}$ then $f(x) = e^x$.*

It is sometimes helpful to represent a function geometrically. For instance, Fig. 1.1 depicts the function f from the set $A = \{p, q, r\}$ to the set $B = \{w, x, y, z\}$ defined by the assignments

$$f(p) = w, \qquad f(q) = y, \qquad f(r) = x.$$

Thus we may consider that a function $f: A \to B$ 'projects' or 'maps' each element of A to its image in B; indeed, functions are frequently called *mappings*.

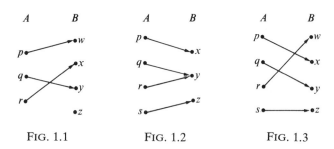

FIG. 1.1   FIG. 1.2   FIG. 1.3

Two functions $f: A \to B$ and $g: A \to B$ with the same domain A and codomain B are called *equal* if $f(x) = g(x)$ for every element x in A.

1.2.2. *Injective, surjective, and bijective functions*

A function $f: A \to B$ is said to be *injective* if, whenever x and x' are distinct elements of A, $f(x)$ and $f(x')$ are distinct elements of B. An injective function from A to B is often called a *one-to-one transformation* of A into B.

A function $f: A \to B$ is called *surjective* when its range is the whole codomain, i.e. for each element $y \in B$ there exists at least one element $x \in A$ such that $f(x) = y$.

Finally, a function $f: A \to B$ is said to be *bijective* if it is both injective and surjective.

Example 1.13. The function $f: P \to \mathbb{N}$ of Example 1.11 is not injective, since $f(x) = f(x')$ for any two people x and x' of the same age. Neither is it surjective, for the set $\mathbb{N}$ contains integers which are not ages of living people.

Example 1.14. The function depicted in Fig. 1.1 is injective but not surjective, whereas the function of Fig. 1.2 is surjective but not injective. Figure 1.3 represents a bijective function.

1.2.3. *Inverse functions*

Any bijective function $f: A \to B$ has the property that, for each element $y \in B$ there exists a unique element $x \in A$ such that $f(x) = y$. (The *existence* of such an element $x \in A$ is assured by the fact that the function is surjective; the *uniqueness* of x follows from the fact that the function is injective.) Hence, it is possible to define a function $f^{-1}: B \to A$ by the rule: *if $y \in B$ then $f^{-1}(y) = x$, where x is the element of A such that $f(x) = y$.* The function $f^{-1}: B \to A$ is called the *inverse* of $f: A \to B$. It will be observed that $f^{-1}: B \to A$ is itself bijective, and that its inverse is $f: A \to B$ again.

Example 1.15. The function $f: \mathbb{R} \to \mathbb{R}^+$, where $f(x) = e^x$ for each $x \in \mathbb{R}$, is bijective. Its inverse $f^{-1}: \mathbb{R}^+ \to \mathbb{R}$ is given by the rule: *if $x \in \mathbb{R}^+$ then $f^{-1}(x) = \ln x$.*

Example 1.16. In some shops the prices of articles are disguised by means of a 'price code', which is a bijective function $f: A \to B$, where $A = \{0, 1, 2, \ldots, 9\}$, B is a set of ten alphabetic characters which, taken in some order, form a *codeword* (such as 'mackintosh'), and the rule of assignment is of the form: if $n \in A$ then $f(n)$ *is the $(n + 1)$th letter of the codeword.* Thus if the codeword used is 'mackintosh', a price of £6.25 is specified on a price tag as t.cn. To interpret this, the shopkeeper uses the assignment rule of the inverse function $f^{-1}: B \to A$.

1.2.4. Composition of functions

Let us suppose that A, B, C, and D are sets and that we have two functions $f: A \to B$ and $g: C \to D$. If $f(A) \subseteq C$, that is if the range of f is contained in the domain of g, then the result of applying first f and then g may be regarded as the application of a single function from A to D. This function, which is called the *composite* of f and g (in that order), and denoted by $g \circ f: A \to D$, is defined by the rule

$$\text{if } x \in A \quad \text{then} \quad (g \circ f)(x) = g(f(x)).$$

Example 1.17. Let $f: \mathbb{R} \to \mathbb{R}$ and $g: \mathbb{R} \to \mathbb{R}$ be two functions defined respectively by the rules: $f(x) = 2x$ *for each* $x \in \mathbb{R}$, and $g(x) = x^2$ *for each* $x \in \mathbb{R}$. Then the function $g \circ f: \mathbb{R} \to \mathbb{R}$ is defined by the rule

$$(g \circ f)(x) = g(f(x)) = 4x^2, \text{ for each } x \in \mathbb{R}$$

while $f \circ g: \mathbb{R} \to \mathbb{R}$ is defined by the rule

$$(f \circ g)(x) = f(g(x)) = 2x^2, \text{ for each } x \in \mathbb{R}.$$

1.3. Binary and *n*-ary operations

1.3.1. Binary operations on a set

If x and y are any two real numbers, then there are various ways of 'operating' on x and y which give another real number. For example, the operation of *addition* gives their *sum* $x + y$, the operation of *multiplication* gives their *product* $x \times y$, and the *subtraction* of y from x gives their *difference* $x - y$. These are all examples of *binary operations* on the set $\mathbb{R}$. Several other binary operations were introduced earlier in this chapter: For instance, if $\mathscr{P}(S)$ is the power set of a given set S, then from any two sets $X, Y \in \mathscr{P}(S)$ we may construct new sets, $X \cup Y$ and $X \cap Y$, which also belong to $\mathscr{P}(S)$; here the operations $\cup$ and $\cap$ are both binary operations on $\mathscr{P}(S)$.

In general terms, if A is any set, then a *binary operation* $\circ$ *on A* is defined as a function $\circ$ from $A \times A$ to A. In other words, a binary operation $\circ$ on a set A is simply a function $\circ$ which assigns, to each ordered pair (x, y) of elements of A, a unique element $\circ(x, y)$ of A.

Since this concept is a generalization of the familiar notions of 'addition' and 'multiplication', it will be helpful to use here the symbolism

$$x \circ y$$

rather than $\circ(x, y)$, for the image of (x, y) under $\circ$.

Example 1.18. We can define a binary operation $\circ$ on the set $S = \{0, 1, 2\}$ by the rule: *if $x, y \in S$ then $x \circ y = max \{x, y\}$*. This operation is represented geometrically in Fig. 1.4. It can also be represented by its *operation table* (Fig. 1.5), where each element $x \circ y$ appears in the cell whose row is labelled 'x' and whose column is labelled 'y'.

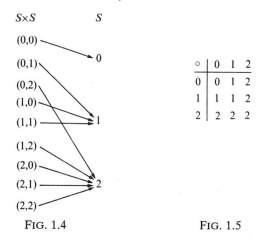

$\circ$	0	1	2
0	0	1	2
1	1	1	2
2	2	2	2

FIG. 1.4   FIG. 1.5

1.3.2. *Idempotent, commutative, and associative operations*

Particular binary operations may possess certain important properties, which are defined below:

Let A be any set. Then an operation $\circ$ on A is

  (i) *idempotent*   if $x \circ x = x$ for all $x \in A$,
 (ii) *commutative* if $x \circ y = y \circ x$ for all $x, y \in A$,
(iii) *associative*   if $(x \circ y) \circ z = x \circ (y \circ z)$ for all $x, y, z \in A$.

Example 1.19. The operations of addition and multiplication on the set of real numbers are both commutative, and both associative, but not idempotent. The operation of subtraction does not have any of the properties (i)–(iii).

Example 1.20. Let $\mathscr{P}(S)$ be the power set of any given set S. Then the operations $\cup$ and $\cap$ on $\mathscr{P}(S)$ both possess all the properties (i)–(iii). (cf. Section 1.1.4).

Example 1.21. The operation defined in Example 1.18 has all the properties (i)–(iii).

Let $\circ$ be any binary operation on a set A. Then, given any number of elements $x_1, x_2, \ldots, x_n$ of A, we define the repeated combination $x_1 \circ x_2 \circ \cdots \circ x_n$, for $n \geq 2$, by the recursive formula

$$x_1 \circ x_2 \circ \cdots \circ x_n = (x_1 \circ x_2 \circ \cdots \circ x_{n-1}) \circ x_n.$$

This determines the arrangement of parentheses (and hence the order of combination) for n elements, given that for $n-1$ elements. As an example, the application of this formula to the case where $n = 4$ gives

$$x_1 \circ x_2 \circ x_3 \circ x_4 = (x_1 \circ x_2 \circ x_3) \circ x_4 = ((x_1 \circ x_2) \circ x_3) \circ x_4$$

Now for $n > 2$, there are obviously other possible ways of arranging the parentheses—for instance, if $n = 4$ there are four other arrangements, viz. $(x_1 \circ (x_2 \circ x_3)) \circ x_4$, $(x_1 \circ x_2) \circ (x_3 \circ x_4)$, $x_1 \circ ((x_2 \circ x_3) \circ x_4)$, and $x_1 \circ (x_2 \circ (x_3 \circ x_4))$—and in general these would all give different results. However, *if the operation $\circ$ is associative, then the value of any combination of elements of A, obtained by repeated application of $\circ$, does not depend on the arrangement of parentheses.* This 'generalized associative law' can be proved by induction, as follows:

First, it follows immediately from the definition (iii) above of associativity that the law is obeyed for all combinations of three elements. To prove that it holds for all combinations of n elements, where $n > 3$, let us assume as an induction hypothesis that the law holds for all combinations of fewer than n elements. Now in whatever order we combine n elements $x_1, x_2, \ldots, x_n$, the last step is to make a combination of the form

$$(x_1 \circ x_2 \circ \cdots \circ x_k) \circ (x_{k+1} \circ \cdots \circ x_n),$$

where k is some integer in the range $1, 2, \ldots, n-1$, the combinations $x_1 \circ x_2 \circ \cdots \circ x_k$ and $x_{k+1} \circ \cdots \circ x_n$ having been calculated previously. By the induction hypothesis, the values of $x_1 \circ x_2 \circ \cdots \circ x_k$ and $x_k \circ \cdots \circ x_n$ do not depend on the way in which their parentheses were arranged. Furthermore, for any value of k in the range $1, 2, \ldots, n-2$,

$$(x_1 \circ x_2 \circ \cdots \circ x_k) \circ (x_{k+1} \circ \cdots \circ x_n)$$
$$= (x_1 \circ x_2 \circ \cdots \circ x_k) \circ (x_{k+1} \circ (x_{k+2} \circ \cdots \circ x_n)),$$

and therefore, since $\circ$ is associative,

$$(x_1 \circ x_2 \circ \cdots \circ x_k) \circ (x_{k+1} \circ \cdots \circ x_n)$$
$$= (x_1 \circ x_2 \circ \cdots \circ x_{k+1}) \circ (x_{k+2} \circ \cdots \circ x_n).$$

It follows that for all values of k in the range $1, 2, \ldots, n-2$,

$$(x_1 \circ x_2 \circ \cdots \circ x_k) \circ (x_{k+1} \circ \cdots \circ x_n) = (x_1 \circ x_2 \circ \cdots \circ x_{n-1}) \circ x_n$$
$$= x_1 \circ x_2 \circ \cdots \circ x_n,$$

as required.

There is also a 'generalized associative and commutative law', which asserts that if an operation $\circ$ on A is both associative and commutative, then the value of any combination of elements of A (obtained by repeated application of $\circ$) is independent of both the order and the grouping of the elements.

1.3.3. *Unit and null elements*

Let A be any set, and let $\circ$ be a binary operation on A. Then any element e of A which has the property

$$x \circ e = x = e \circ x, \quad \text{for all } x \in A,$$

is called a *unit element* (or a *neutral element*) for the operation $\circ$.

For any particular binary operation $\circ$ on a set A, the set A does not necessarily contain a unit element; but if A does contain such an element, it is unique. For suppose that A contains two distinct unit elements e and e' for $\circ$; then putting $x = e'$ in the equation $x \circ e = x = e \circ x$ gives

$$e' \circ e = e' = e \circ e',$$

while replacing e by e' and putting $x = e$ gives

$$e \circ e' = e = e' \circ e,$$

whence $e = e'$, contradicting our initial assumption.

Example 1.22. The number 1 is a unit element for ordinary multiplication, since $1 \times x = x = x \times 1$ for all $x \in \mathbb{R}$. The number 0 is a unit element for ordinary addition, since $0 + x = x = x + 0$, for all $x \in \mathbb{R}$.

Example 1.23. *Languages over an alphabet.* This example introduces some concepts of the algebraic theory of languages, which will be applied later to the enumeration of paths on graphs.

We call any finite set of symbols an *alphabet*, and describe the elements of an alphabet as its *letters*. A *word* (or *string*) *over an alphabet* Σ is a finite sequence of zero or more letters of Σ. The sequence of zero letters is called the *empty word*, and is denoted by λ. Thus

$$\lambda, \text{a, cab, dada, baaa}$$

are all words over the alphabet {a, b, c, d}. The set of all words over an alphabet Σ is denoted by Σ^*, and the subsets of Σ^* are called *languages over the alphabet* Σ.

If P and Q are words over an alphabet Σ, then the sequence of letters obtained by concatenating or 'linking together' P and Q is also a word over Σ. Hence concatenation is a binary operation on Σ^*. Obviously, concatenation is not commutative: For instance, denoting concatenation on the English alphabet by $\circ$, we have

$$\text{se} \circ \text{ver} = \text{sever} \quad \text{whereas} \quad \text{ver} \circ \text{se} = \text{verse}.$$

However, it is evident that concatenation is associative:

$$(\text{joy} \circ \text{ful}) \circ \text{ness} = \text{joy} \circ (\text{ful} \circ \text{ness}) = \text{joyfulness}.$$

It will also be noted that for any word P of Σ^*,

$$P \circ \lambda = P = \lambda \circ P.$$

Thus λ is the unit element for concatenation on Σ^*.

Now let $\circ$ be a binary operation on a set A, and let us suppose that A has a unit element e for the operation $\circ$. Then it may happen that a given element $x \in A$ has an *inverse* with respect to $\circ$, that is to say, there may exist an element $\tilde{x} \in A$ such that

$$x \circ \tilde{x} = e = \tilde{x} \circ x.$$

It is easily proved that if the operation $\circ$ is associative then the inverse of any given element x, when it exists, is unique. (We leave the proof as an exercise for the reader.)

Example 1.24. With respect to the operation of addition on $\mathbb{R}$, every number $x \in \mathbb{R}$ has an inverse, which is $-x$. With respect to multiplication, every number $x \in \mathbb{R}$ other than 0 has an inverse, which is $1/x$.

Given a set A and a binary operation $\circ$ on A, we call any element n of A which has the property

$$x \circ n = n = n \circ x, \quad \text{for all } x \in A,$$

a *null element* (or *zero element*) *for the operation* $\circ$. As with unit elements, a given binary operation does not necessarily have a null element, but if it does have one, then this is unique.

Example 1.25. The number 0 is a null element for ordinary multiplication, since $0 \times x = 0 = 0 \times x$ for all $x \in \mathbb{R}$; there is no null element for ordinary addition.

Example 1.26. For the operation of Example 1.18, the number 0 is a unit element, while the number 2 is a null element.

Example 1.27. For the operation $\cup$ on the power set $\mathscr{P}(S)$ of a given set S, the unit and null elements are ϕ and S respectively, since $X \cup \phi = X$ and $X \cup S = S$, for all $X \in \mathscr{P}(S)$. For the operation $\cap$ these roles are reversed, i.e. ϕ is the null element and S is the unit element.

1.3.4. Cancellation

Let A be a set equipped with a binary operation $\circ$. Then an element $x \in A$ is said to be *left-cancellative* (with respect to $\circ$) if

$$x \circ y = x \circ z \quad \text{implies that} \quad y = z$$

or, to give an equivalent condition, if

$$y \neq z \quad \text{implies that} \quad x \circ y \neq x \circ z.$$

Similarly, an element x is *right-cancellative* if

$$y \circ x = z \circ x \quad \text{implies that} \quad y = z.$$

An element which is both left- and right-cancellative is described simply as being *cancellative*. If every element of A is cancellative, other than the null element for $\circ$ (if this exists), we say that the operation $\circ$ has the *cancellation property*.

Example 1.28. On the set $\mathbb{R}$ of real numbers, the operations of addition and multiplication both have the cancellation property, but the operation $x \circ y = \max{(x, y)}$ does not possess this property.

1.3.5. Distributivity

Let $*$ and $\circ$ be two binary operations on a set A. Then the operation $*$ is *distributive over* $\circ$ *from the left* if

$$x * (y \circ z) = (x * y) \circ (x * z), \quad \text{for all } x, y, z \in A, \qquad (1)$$

and it is *distributive over* $\circ$ *from the right* if

$$(y \circ z) * x = (y * x) \circ (z * x), \quad \text{for all } x, y, z \in A. \qquad (2)$$

If both the laws (1) and (2) are obeyed, we say simply that the operation $*$ is *distributive over the operation* $\circ$.

It is evident that, if the operation $*$ is commutative, each of the distributive laws (1) and (2) implies the other.

Example 1.29. On the power set $\mathcal{P}(A)$ of any set A, the operations of union and intersection are each distributive over the other.

Now let us suppose that a set A is equipped with two binary operations $*$ and $\circ$, where $*$ is distributive over $\circ$, and the operation $\circ$ is associative. Then, from the distributive law (1) above we obtain the n-term left-distributive law

$$x * (y_1 \circ y_2 \circ \cdots \circ y_n) = (x * y_1) \circ (x * y_2) \circ \cdots \circ (x * y_n), \qquad (1')$$

and from (2) above we obtain the m-term right-distributive law

$$(x_1 \circ x_2 \circ \cdots \circ x_m) * y = (x_1 * y) \circ (x_2 * y) \circ \cdots \circ (x_m * y). \qquad (2')$$

The formulae (1') and (2') are easily proved, by induction on n and m. Combining these formulae, we get the *generalized distributive law*:

$$(x_1 \circ \cdots \circ x_m) * (y_1 \circ \cdots \circ y_n) = (x_1 * y_1) \circ \cdots \circ (x_1 * y_n) \circ$$

$$\cdots \circ (x_m * y_1) \circ \cdots \circ (x_m * y_n).$$

1.3.6. n-ary operations

In Section 1.3.1 we defined a binary operation on a set A as a function from A^2 to A (where A^2 denotes the Cartesian product $A \times A$). More generally, given a set A and an integer n, we describe any function from A^n to A as an *n-ary operation* on A. For $n = 1$ we have a *unary operation*, which is just a function $f: A \to A$ in the usual sense; for instance the square-root operation is a unary operation on $\mathbb{R}^+$.

When we have several operations (of any kind) defined on a set A, we may describe the collection of these operations as an *algebraic structure* on A; the set A with this structure is also called an *algebra*. Some particular types of algebras will be presented in Section 1.5.

Exercises

1.5. Test the following operations $a \circ b$ on real numbers a and b for being idempotent, commutative, and associative:

   (i) $a \circ b = (a + b)/2$,
   (ii) $a \circ b = (a \times b)/2$,
   (iii) $a \circ b = a$,
   (iv) $a \circ b = \min\{a, b\}$,
   (v) $a \circ b = a + b - (a \times b)$.

1.6. Let $\circ$ be a binary operation on a set A. Prove that if an element x is both a unit element and a null element for $\circ$, then $A = \{x\}$.

1.7. The table below defines a binary operation $\circ$ on the set $S = \{\bigcirc, \square, \triangle\}$. By inspection of the table, determine whether this operation is idempotent, and whether it is commutative. Does S contain a unit element for $\circ$? If so, which elements of S are invertible, and what are their inverses? Does S contain a null element for $\circ$? Does the operation $\circ$ have the cancellation property?

$\circ$	$\bigcirc$	$\square$	$\triangle$
$\bigcirc$	$\bigcirc$	$\square$	$\triangle$
$\square$	$\square$	$\triangle$	$\bigcirc$
$\triangle$	$\triangle$	$\bigcirc$	$\square$

1.4. Binary Relations

1.4.1. The concept of relation

It frequently happens that two objects are 'related' or associated with each other in some way. For instance, we may say of two people x and y that 'x is the mother of y', or 'x is older than y'. And if x and *y* are positive integers, we may have that '*x* is a multiple of *y*' or '$x \leq y$'. Each of these statements is said to express a *relation* between two objects.

In the following general discussion, the symbol $\mathcal{R}$ will be used to denote an arbitrary relation between two objects (thus '$\mathcal{R}$' may stand for 'is the mother of', or 'is older than', or 'is a multiple of', etc.), and we shall write

$$x \, \mathcal{R} \, y$$

to indicate that *x* stands in the relation $\mathcal{R}$ to *y*.

Now let *A* and *B* be any sets. Then $\mathcal{R}$ is called a *binary relation from A to B* if for any given pair (x, y) of $A \times B$, the condition $x \, \mathcal{R} \, y$ either does, or does not, hold. (In other words, $\mathcal{R}$ is a binary relation from *A* to *B* if, for each pair $(x, y) \in A \times B$, the statement '$x \, \mathcal{R} \, y$' is *meaningful*, being either true or false for that particular pair.) In the particular case where $A = B$, we describe a relation $\mathcal{R}$ from *A* to *B* as a *relation on A*.

Example 1.30. Let *P* be the set of all living people, and let *M* be the set of months of the year. Then 'was born in the month of' denotes a binary relation from *P* to *M*, while the expression 'is a parent of' denotes a relation on *P*.

The concept of a binary relation from a set *A* to a set *B* is closely connected with that of a function from *A* to *B*. Indeed, each function $f: A \to B$ determines a binary relation $\mathcal{R}$ from *A* to *B*, through the rule

$$x \, \mathcal{R} \, y \quad \text{if and only if} \quad y = f(x).$$

Conversely if $\mathcal{R}$ is a given relation from *A* to *B*, *such that for each* $x \in A$, *there is precisely one* $y \in B$ *with* $x \, \mathcal{R} \, y$, then $\mathcal{R}$ determines a function $f: A \to B$, through the rule

$$y = f(x) \quad \text{if and only if} \quad x \, \mathcal{R} \, y.$$

In this sense, the concept of a relation is a generalization of that of a function.

1.4.2. *Complementary and converse relations*

For any relation $\mathcal{R}$ from a set *A* to a set *B* there exists a *complementary relation* or *negation* $\bar{\mathcal{R}}$, such that for each pair

$(x, y) \in A \times B$, $x \bar{\mathcal{R}} y$ holds if and only if $x \mathcal{R} y$ does *not* hold. Also, for any relation $\mathcal{R}$ from A to B we can define a *converse relation* $\mathcal{R}'$ from B to A, by the rule

$$y \mathcal{R}' x \quad \text{if and only if} \quad x \mathcal{R} y, \quad \text{for all } (x, y) \in A \times B.$$

Example 1.31. The relation 'is a parent of' on the set of living people has the complement 'is not a parent of', while its converse is the relation 'is a child of'.

1.4.3. *Some special kinds of relations on a set*

Let A be any set, and let $\mathcal{R}$ be a binary relation on A. Then $\mathcal{R}$ is said to be

Reflexive   when $x \mathcal{R} x$ for all $x \in A$;

Anti-reflexive   when $x \mathcal{R} x$ for no $x \in A$;

Symmetric   when $x \mathcal{R} y$ implies $y \mathcal{R} x$;

Anti-symmetric   when $x \mathcal{R} y$ and $y \mathcal{R} x$ together imply $x = y$;

Transitive   when $x \mathcal{R} y$ and $y \mathcal{R} z$ together imply $x \mathcal{R} z$.

Example 1.32. The relation 'has the same parents as' on the set of living people is reflexive, symmetric, and transitive.

Example 1.33. The relation $\leq$ on $\mathbb{R}$ is reflexive, anti-symmetric, and transitive, while the relation $<$ is *anti*-reflexive, anti-symmetric, and transitive.

A relation on a set A is called a *pre-ordering* of A when it is reflexive and transitive. A pre-ordering which is symmetric is called an *equivalence relation*; whereas an anti-symmetric pre-ordering is called an *ordering* of A.

Thus in Example 1.32 the relation 'has the same parents as' is an equivalence relation, whereas in Example 1.33 the relation $\leq$ is an ordering of $\mathbb{R}$. These two types of relations, which are of particular importance, are considered in more detail in the following sections.

1.4.4. *Equivalence relations*

A relation on a set A is called an *equivalence relation on A* when it is reflexive, symmetric, and transitive.

There is an important connection between equivalence relations and partitions, which were introduced in Section 1.1.7. Let $\mathscr{R}$ be an equivalence relation on a set A, and for each element $x \in A$ let us define the set

$$E_x = \{y \in A \mid x \mathscr{R} y\},$$

called the *equivalence class of x* (*with respect to $\mathscr{R}$*). Note that since $\mathscr{R}$ is reflexive, $x \in E_x$, for each $x \in A$.

Now for each pair of elements x and y of A, the equivalence classes E_x and E_y are either equal or disjoint, with

(i) $E_x = E_y$,   if $x \mathscr{R} y$,

(ii) $E_x \cap E_y = \phi$,   if $x \bar{\mathscr{R}} y$.

This can be proved as follows:

(i) Suppose $x \mathscr{R} y$. Then, for each element $z \in E_y$, $x \mathscr{R} y$, and $y \mathscr{R} z$, and therefore (by transitivity) $x \mathscr{R} z$, which implies that $z \in E_x$; hence $E_y \subseteq E_x$. Also, by symmetry $y \mathscr{R} x$, and reversing the roles of x and y in the above argument gives $E_x \subseteq E_y$. Combining these results, we obtain $E_x = E_y$.

(ii) Alternatively, suppose $x \bar{\mathscr{R}} y$. Then E_x and E_y cannot have any elements in common; for if an element z belonged to both E_x and E_y, it would follow that $x \mathscr{R} z$ and $y \mathscr{R} z$, and hence (by symmetry) that $x \mathscr{R} z$ and $z \mathscr{R} y$, and therefore (by transitivity) that $x \mathscr{R} y$; but this would be contrary to our initial assumption that $x \bar{\mathscr{R}} y$.

Since every element of A belongs to some equivalence class, and distinct equivalence classes are disjoint, the set of all equivalence classes is a partition of A (as defined in Section 1.1.8). This partition is called the *partition of A induced by $\mathscr{R}$*, and is denoted by $A/\mathscr{R}$.

Conversely, given any partition $\mathscr{S}$ of a set A we may define a relation $\mathscr{R}$ on A by the rule: $x \mathscr{R} y$ if and only if x and y belong to the same member of $\mathscr{S}$. It is easy to verify that this is an equivalence relation on A, and that the equivalence classes with respect to $\mathscr{R}$ are the sets which are members of $\mathscr{S}$.

Example 1.34. Let P be the set of all living people, and let $\mathscr{R}$ be the equivalence relation 'was born under the same sign of the Zodiac as'. Then for any person x, the equivalence class of x (with respect to $\mathscr{R}$) is the set of all people born under the same sign as x. The twelve distinct sets of this kind constitute a partition of P.

1.4.5. Orderings

A relation on a set A is called an *ordering* of A when it is reflexive, anti-symmetric, and transitive. Orderings are often indicated by the special relation symbol $\leqslant$ (the notation $x \leqslant y$ is read as 'x is inferior or equal to y'). Using this notation, the characteristic properties of an ordering can be described as follows:

(i) $x \leqslant x$, for all $x \in A$;   (reflexivity)

(ii) if $x \leqslant y$ and $y \leqslant x$, then $x = y$;   (anti-symmetry)

(iii) if $x \leqslant y$ and $y \leqslant z$, then $x \leqslant z$.   (transitivity)

Two elements x and y of A are said to be *comparable* if $x \leqslant y$ or $y \leqslant x$; otherwise they are incomparable. If all the elements of A taken two at a time are comparable, the ordering $\leqslant$ of A is *total*, and A is said to be a *totally ordered set* or *chain*.

Example 1.35. Let A be any set. Then the inclusion relation $\subseteq$ on $\mathscr{P}(A)$ is an ordering of $\mathscr{P}(A)$, since (i) $X \subseteq X$, for all $X \in \mathscr{P}(A)$, (ii) if $X \subseteq Y$ and $Y \subseteq X$, then $X = Y$, and (iii) if $X \subseteq Y$ and $Y \subseteq Z$, then $X \subseteq Z$. However, if A contains more than one element, this ordering $\subseteq$ of $\mathscr{P}(A)$ is not total: For if A contains two distinct elements x and y say, then $\mathscr{P}(A)$ contains the two distinct elements $\{x\}$ and $\{y\}$, and these are incomparable, i.e. neither $\{x\} \subseteq \{y\}$ nor $\{y\} \subseteq \{x\}$ holds.

Example 1.36. The ordering $\leq$ of the set of real numbers is total, since for any real numbers x and y, either $x \leq y$ or $y \leq x$.

It will be convenient to adopt the following conventions: $x \geqslant y$ (read as 'x is superior or equal to y') has the same significance as $y \leqslant x$; $x < y$ (read as 'x is inferior to y') means $x \leqslant y$ but $x \neq y$; and $x > y$ (read as 'x is superior to y') means $x \geqslant y$ but $x \neq y$.

An ordered set having only a few elements can be depicted conveniently by a *Hasse diagram*, in which each element is represented by a point, so placed that if $x < y$ then the point representing x lies below the point representing y; lines are drawn to connect elements x, y such that y *covers* x, i.e. $y > x$ and there is no element z such that $y > z > x$. For example, Fig. 1.6 represents the power set $\mathscr{P}(S)$ of a set $S = \{a, b, c\}$, ordered by the set inclusion relation.

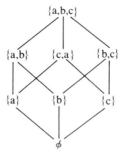

FIG. 1.6

Least and greatest elements. If an ordered set A contains an element ϕ such that $\phi \leqslant x$ for all $x \in A$, then ϕ is called the *least element* of A. A set can have at most one least element: For if ϕ and ϕ' were two least elements, we would have $\phi \leqslant \phi'$ and $\phi' \leqslant \phi$, hence (by the anti-symmetry rule (ii) above) $\phi = \phi'$.

If A contains an element u such that $x \leqslant u$ for all $x \in A$, then u is the *greatest element* of A. A set can have at most one greatest element.

Minimal and maximal elements. An element m of a set A is a *minimal element* of A if there does not exist any element in A which is strictly inferior to m:

$$x \nleqslant m \quad \text{for all } x \in A.$$

In any set with a least element ϕ, clearly ϕ is the only minimal element. However, a minimal element, if it exists, is not necessarily unique: For instance, the set $\{\{a\}, \{b\}, \{a, b\}\}$ ordered by the inclusion relation $\subseteq$ has two minimal elements, $\{a\}$ and $\{b\}$.

In the same way, an element m of a set A is a *maximal* element of A if there does not exist any element in A which is strictly superior to m.

Greatest lower and least upper bounds. Let B be any subset of an ordered set A. Then we call an element $p \in A$ a *lower bound* of B if $p \leqslant x$ for all $x \in B$; similarly, we call $p \in A$ an *upper bound* of B if $p \geqslant x$ for all $x \in B$. If the set of lower bounds of B has a greatest element, this is called the *greatest lower bound* (g.l.b.) of B; similarly, if the set of upper bounds of B has a least element, this is called the *least upper bound* (l.u.b.) of B.

Example 1.37. The relation $x \mid y$ (meaning that x is a factor of y) is an ordering of the set $\mathbb{J}$ of all positive integers. With respect to this ordering, $\mathbb{J}$ has a least element (the integer 1). For any subset B of $\mathbb{J}$, any common factor of the integers in B is a lower bound of B; the g.l.b. of B is just the highest common factor of the integers in B.

Exercises

1.8. Test the following relations on the set of living people for being reflexive, anti-reflexive, symmetric, anti-symmetric, and transitive:

   (i) is taller than
   (ii) lives within a mile of
   (iii) is married to.

1.9. Let $\leqslant$ be an ordering on a set A. Prove that in A, x and y are equal if and only if, for any element $z \in A$, $z \leqslant x$ implies $z \leqslant y$ and conversely.

1.10. Let Σ be an alphabet, with a total ordering $\leqslant$. Let Σ^* be the set of all words on Σ, and let α be the relation on Σ^* defined by the rule: for any two words $X = x_1 x_2 \ldots x_m$ and $Y = y_1 y_2 \ldots y_n$ of Σ^*, $X\alpha Y$ if either (i) $m \leq n$, and $x_i = y_i$ for $1 \leq i \leq m$, or (ii) $x_i = y_i$ for $1 \leq i < k$, and $x_k < y_k$, for some $k \leq m, n$.
   Prove that the relation α is a total ordering of Σ^*. (The ordering α is called the *lexicographic ordering* of Σ^* associated with the *alphabetic ordering* $\leqslant$ of Σ.)

1.5. Lattices

1.5.1. Introduction

We define a *join-semilattice* as an ordered set in which any two elements x and y have a least upper bound; we call this bound the *join* of x and y, and denote it by $x \vee y$.

In a similar fashion, we define a *meet-semilattice* as an ordered set in which any two elements x and y have a greatest lower bound; this bound is called the *meet* of x and y, and it is denoted by $x \wedge y$.

A *lattice* is an ordered set which is both a join-semilattice and a meet-semilattice. In other words, a lattice is an ordered set in which every pair of elements has a least upper bound and a greatest lower bound.

Example 1.38. The power set $\mathscr{P}(A)$ of a given set A, ordered by the inclusion relation $\subseteq$, is a lattice. For all $X, Y \in \mathscr{P}(A)$, $X \vee Y = X \cup Y$ and $X \wedge Y = X \cap Y$.

Example 1.39. Any chain is a lattice, in which $x \vee y$ is simply the greater and $x \wedge y$ the lesser of x and y. One example is the set of real numbers with the usual ordering $\le$, where $x \vee y = \max\{x, y\}$ and $x \wedge y = \min\{x, y\}$.

Example 1.40. Of the four ordered sets whose Hasse diagrams are shown in Fig. 1.7, the first three are lattices. The fourth is a join-semilattice, but it is not a meet-semilattice since the elements c and d do not have a greatest lower bound.

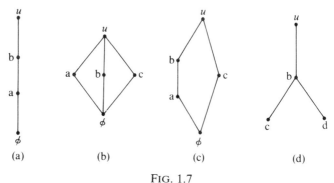

FIG. 1.7

1.5.2. Algebraic definition of a semilattice and lattice

In the previous section we presented semilattices and lattices in terms of an order relation, but alternatively, we can regard a semilattice or lattice as a set equipped with one, or two, binary operations respectively.

For instance, in a join-semilattice we may consider the formation of the element $x \vee y$ from x and y as a binary operation defined on the semilattice. It is easily verified that this operation obeys the following laws:

$$x \vee x = x \qquad \text{(Idempotent law)}$$

$$x \vee y = y \vee x \qquad \text{(Commutative law)}$$

$$(x \vee y) \vee z = x \vee (y \vee z) \quad \text{(Associative law)}.$$

Indeed, the idempotent and commutative laws follow immediately from the definition of a least upper bound; and since the l.u.b. of the set $\{x, y, z\}$ can be written either as $(x \vee y) \vee z$ or $x \vee (y \vee z)$, the associative law also holds.

Conversely, let us consider a set A, equipped with a binary operation $\circ$ which is idempotent, commutative, and associative. Let us define a relation $\mathscr{R}$ on A through the rule

$$x \mathscr{R} y \quad \text{if and only if} \quad x \circ y = y.$$

Then,

(i) since the operation $\circ$ is idempotent, $\mathscr{R}$ is *reflexive*: $x \circ x = x$ implies $x \mathscr{R} x$.

(ii) $\mathscr{R}$ is *anti-symmetric*: the conditions $x \mathscr{R} y$ and $y \mathscr{R} x$ are equivalent to $x \circ y = y$ and $y \circ x = x$, from which it follows (by the commutativity of $\circ$) that $x = y$;

(iii) $\mathscr{R}$ is *transitive*: the conditions $x \mathscr{R} y$ and $y \mathscr{R} z$ are equivalent to $x \circ y = y$ and $y \circ z = z$, from which it follows (by the associativity of $\circ$) that $x \circ z = x \circ (y \circ z) = (x \circ y) \circ z = y \circ z = z$, which implies that $x \mathscr{R} z$.

Thus $\mathscr{R}$ *is an ordering* of A.

Furthermore, *the set A with ordering $\mathscr{R}$ is a join-semilattice, $x \circ y$ being the join of x and y*. Indeed, $x \mathscr{R} (x \circ y)$ since

$$x \circ (x \circ y) = (x \circ x) \circ y = x \circ y,$$

and $y \mathscr{R} (x \circ y)$ since

$$y \circ (x \circ y) = (x \circ y) \circ y = x \circ (y \circ y) = x \circ y,$$

and therefore $x \circ y$ is an upper bound for x and y. If z is any other upper bound of x and y then $x \mathscr{R} z$ and $y \mathscr{R} z$, hence $x \circ z = z$ and $y \circ z = z$, and therefore

$$(x \circ y) \circ z = x \circ (y \circ z) = x \circ z = z,$$

which implies that $(x \circ y) \mathscr{R} z$. It follows that $x \circ y$ is the least upper bound of x and y, which proves our assertion.

It would have been possible, in place of the relation $\mathscr{R}$, to consider its converse $\mathscr{R}'$, which can be defined directly in terms of the $\circ$ operation through the rule

$$x \mathscr{R}' y \quad \text{if and only if} \quad x \circ y = x.$$

It is evident that $\mathscr{R}'$ is also an ordering of A, and by a similar argument to our previous one we can show that *the set A with ordering $\mathscr{R}'$ is a meet-semilattice, $x \circ y$ being the meet of x and y*.

Thus, the set A, equipped with the operation ∘, can be regarded both as a join-semilattice and as a meet-semilattice. This justifies the following 'algebraic' definition:

A *semilattice* is a set with a binary operation which is idempotent, commutative and associative.

Using similar arguments, we can obtain an algebraic definition of a lattice by regarding the formation of the join $x \vee y$ and the meet $x \wedge y$ of x and y as two binary operations. By the definitions of the join and meet, the operations $\vee$ and $\wedge$ obey the following laws:

L1 $x \vee x = x$ *and* $x \wedge x = x$ (*Idempotent laws*)

L2 $x \vee y = y \vee x$ *and* $x \wedge y = y \wedge x$ (*Commutative laws*)

L3 $(x \vee y) \vee z$ *and* $(x \wedge y) \wedge z$ (*Associative laws*)
       $= x \vee (y \vee z)$ $= x \wedge (y \wedge z)$

L4 $x \wedge (x \vee y) = x$ *and* $x \vee (x \wedge y) = x$ (*Absorption laws*)

The validity of L1–L3 has already been established; the first part of L4 follows from the fact that $x \leqslant x \vee y$, while the second part results from the fact that $x \geqslant x \wedge y$.

In the algebraic description we start from these properties, and we define a lattice as a set equipped with two binary operations, denoted by $\vee$ and $\wedge$, which obey the laws L1–L4. The equivalence of this definition with that given in the previous section is established by the following theorem:

Let L be a set with two binary operations $\vee$ and $\wedge$ satisfying L1–L4. Then an ordering $\leqslant$ may be defined on L, by the rule

L5 $x \leqslant y$ *if and only if* $x \vee y = y$,

and relative to this ordering L is a lattice (as defined in Section 1.5.1) in which $x \vee y$ and $x \wedge y$ are respectively the join and meet of x and y.

To prove the theorem, we first observe that

L6 $x \vee y = y$ *if and only if* $x \wedge y = x$
                                         (*Consistency law*)

For if $x \vee y = y$ then (by L4) $x = x \wedge (x \vee y) = x \wedge y$, and the converse holds by symmetry. Next, it follows from our discussion of semilattices that since the $\vee$ operation on L is idempotent, commutative, and associative, the relation $\leqslant$ defined by L5 is indeed an ordering, and relative to this ordering L is a join-semilattice in which $x \vee y$ is the join of x and y. Similarly, since the $\wedge$ operation is idempotent, commutative, and associative, and (by L5 and L6) the ordering $\leqslant$ can be expressed as

L7 $\qquad\qquad x \leqslant y \quad if\ and\ only\ if \quad x \wedge y = x,$

the set L with the ordering $\leqslant$ is a meet-semilattice in which $x \wedge y$ is the meet of x and y, which completes the proof.

1.5.3. The principle of duality for lattices

If in the identities L1–L4 we interchange the symbols $\vee$ and $\wedge$ throughout, the identities are preserved. Moreover, when we interchange $\vee$ and $\wedge$, the condition $x \vee y = y$ becomes $x \wedge y = y$ which (by L7) is equivalent to $x \vee y = x$; thus (by L5) the condition $x \leqslant y$ becomes $y \leqslant x$, which can also be written as $x \geqslant y$. This suggests and essentially proves the following *principle of duality*:

Every statement which is deducible from the laws L1–L4 and the definition L5 of the relation $\leqslant$ remains valid if the operation symbols $\vee$ and $\wedge$ and the relation symbols $\leqslant$ and $\geqslant$ are interchanged throughout the statement.

1.5.4. Some further properties of lattices

(i) If a lattice L contains a least element then, by L5 and L7,

L8 $\qquad\quad \phi \vee x = x \quad and \quad \phi \wedge x = \phi \quad for\ all\ x \in L;$

thus the least element of L is the unit element for $\vee$ and the null element for $\wedge$. Similarly, if a lattice L has a greatest element u, then

L9 $\qquad\quad u \vee x = u \quad and \quad u \wedge x = x \quad for\ all\ x \in L,$

so u is the null element for $\vee$ and the unit element for $\wedge$.

(ii) Every lattice has the following property:

L10 $\quad$ *The conditions $x \leqslant z$ and $y \leqslant z$ are together equivalent to $x \vee y \leqslant z$.*

For if $x \leqslant z$ and $y \leqslant z$ then (by L5) $x \vee z = z$ and $y \vee z = z$, hence $(x \vee y) \vee z = x \vee (y \vee z) = x \vee z = z$, which implies that $x \vee y \leqslant z$; conversely, if $x \vee y \leqslant z$ then (since $x \leqslant x \vee y$ and $y \leqslant x \vee y$), $x \leqslant z$ and $y \leqslant z$. Applying the duality principle to L10, we obtain

L10′   *The conditions* $z \leqslant x$ *and* $z \leqslant y$ *are together equivalent to* $z \leqslant x \wedge y$.

(iii) In any lattice L,

L11 $x \leqslant y$ *implies* $x \vee z \leqslant y \vee z$ *and* $x \wedge z \leqslant y \wedge z$, *for all* $z \in L$.

Indeed, if $x \leqslant y$ then $x \vee y = y$, and therefore $(x \vee z) \vee (y \vee z) = (x \vee y) \vee (z \vee z) = y \vee z$, which implies that $x \vee z \leqslant y \vee z$; that $x \wedge z \leqslant y \wedge z$ follows by duality. The property L11 is expressed in words by saying that the operations $\vee$ and $\wedge$ are *isotone* for the ordering $\leqslant$.

1.5.5. Distributive lattices and complemented lattices

A lattice L is said to be *distributive* if it satisfies the laws

L12 $x \wedge (y \vee z) = (x \wedge y) \vee (x \wedge z)$ *for all* $x, y, z \in L$,

L12′ $x \vee (y \wedge z) = (x \vee y) \wedge (x \vee z)$ *for all* $x, y, z \in L$.

Actually these two laws are not independent: each implies the other. To show that L12 implies L12′ we have from L12 that

$$(x \vee y) \wedge (x \vee z) = [(x \vee y) \wedge x] \vee [(x \vee y) \wedge z]$$

$$= x \vee [(x \vee y) \wedge z] \qquad \text{(by L2 and L4)}$$

$$= x \vee [(x \wedge z) \vee (y \wedge z)] \qquad \text{(by L2 and L12)}$$

$$= [x \vee (x \wedge z)] \vee (y \wedge z) \qquad \text{(by L3)}$$

$$= x \vee (y \wedge z) \qquad \text{(by L4)}.$$

By duality, it follows that L12′ implies L12.

Example 1.41. Any chain is a distributive lattice. For in a chain $x \wedge y$ is the lesser of x and y, while $x \vee y$ is the greater of x and y. Thus $x \wedge (y \vee z)$ and $(x \wedge y) \vee (x \wedge z)$ are both equal to x if x is inferior to y or z; and both are equal to $y \vee z$ in the alternative case where x is superior to y and z.

Now let us suppose that in a distributive lattice, three elements x, y, and z satisfy the conditions

$$z \vee x = z \vee y \quad \text{and} \quad z \wedge x = z \wedge y.$$

Then, using the absorptive, commutative and distributive laws we obtain

$$x = x \vee (z \wedge x) = x \vee (z \wedge y) = (x \vee z) \wedge (x \vee y)$$

$$= (y \vee z) \wedge (y \vee x) = y \vee (z \wedge x) = y \vee (z \wedge y) = y.$$

Thus, in any distributive lattice,

L13 $z \vee x = z \vee y$ *and* $z \wedge x = z \wedge y$ *together imply* $x = y$.

Complements. In a lattice with a least element ϕ and a greatest element u, two elements x and y are said to be *complementary* whenever

$$x \wedge y = \phi \quad \text{and} \quad x \vee y = u.$$

An element which is complementary to x is also called a *complement* of x.

Example 1.42. In any lattice L with a least element ϕ and a greatest element u, the element ϕ is the unique complement of u, and u is the unique complement of ϕ. However in general, the elements of L need not all have complements; for instance, in the lattice of Fig. 1.7(a), the elements a and b do not have complements. Alternatively, an element may have several complements; for example, in Fig. 1.7(b) each of the elements a, b, c is complementary to the other two.

Example 1.43. Consider the lattice formed by the power set $\mathscr{P}(A)$ of a given set A, the operations $\vee$ and $\wedge$ being set union and intersection respectively. The least and greatest elements of this lattice are the null set ϕ and the set A respectively. Each element $X \in \mathscr{P}(A)$ has exactly one complement, which is the set difference $A - X$ (i.e. the complement of X in A in the set-theoretic sense).

It will be observed that *in a distributive lattice, a given element x can have at most one complement.* For suppose that in such a lattice, two elements y and z satisfy the conditions

$$x \wedge y = \phi \quad \text{and} \quad x \vee y = u,$$

$$x \wedge z = \phi \quad \text{and} \quad x \vee z = u.$$

Then

$$x \wedge y = x \wedge z \quad \text{and} \quad x \vee y = x \vee z,$$

and by L13 these conditions together imply that $y = z$.

In a distributive lattice, the unique complement of a given element x, when it exists, is usually denoted by $\bar{x}$.

A *complemented lattice* is a lattice with a least and greatest element, in which every element has at least one complement.

Clearly, if a lattice is distributive and complemented, each of its elements has a unique complement. The lattices with these particular properties are considered in more detail below.

1.5.6. Boolean algebras

A lattice which is distributive and complemented is called a *Boolean lattice*, or *Boolean algebra*.

Example 1.44. The power set $\mathscr{P}(A)$ of a given set A, with binary operations $\cup$ and $\cap$, forms a Boolean algebra. (See Example 1.38.)

Example 1.45. The lattice comprising two elements x and y with $x < y$ is distributive and complemented, with $\bar{x} = y$ and $\bar{y} = x$. It is called the *two-element Boolean algebra*.

The two elements x and y of this algebra are often denoted by the numerals 0 and 1. In this case the ordering $\leqslant$ can be interpreted as the usual ordering $\leq$ of integers, and the operations $\vee$ and $\wedge$ take the significance

$$x \vee y = \max\{x, y\}, \qquad x \wedge y = \min\{x, y\},$$

or in terms of the usual arithmetic operations,

$$x \vee y = x + y - (x \times y), \qquad x \wedge y = x \times y.$$

Clearly, since a Boolean algebra is distributive and complemented, all the properties of distributive lattices and complemented lattices are properties of a Boolean algebra. Moreover, in a Boolean algebra we have

L14 $$\bar{\bar{x}} = x \qquad \qquad \textit{(law of involution)},$$

L15 $$\overline{(x \vee y)} = \bar{x} \wedge \bar{y} \quad and \quad \overline{(x \wedge y)} = \bar{x} \vee \bar{y}$$
$$\textit{(de Morgan's laws)}.$$

The law L14 follows immediately from the unicity of complements. To prove the first part of L15 we observe that

$$(x \vee y) \wedge (\bar{x} \wedge \bar{y}) = (x \wedge \bar{x} \wedge \bar{y}) \vee (y \wedge \bar{x} \wedge \bar{y})$$
$$= (\phi \wedge \bar{y}) \vee (\phi \wedge \bar{x}) = \phi \wedge \phi = \phi$$

and

$$(x \vee y) \vee (\bar{x} \wedge \bar{y}) = (x \vee y \vee \bar{x}) \wedge (x \vee y \vee \bar{y})$$
$$= (u \vee y) \wedge (u \vee x) = u \vee u = u,$$

from which it follows that $\bar{x} \wedge \bar{y}$ is the complement of $x \vee y$. By a dual argument, $\bar{x} \vee \bar{y}$ is the complement of $x \wedge y$.

Exercises

1.11. Prove that if the binary operations $\vee$ and $\wedge$ on a set L obey the laws L2–L4, then they obey the law L1.

1.12. Prove that in any lattice, the conditions $w \leqslant x$ and $y \leqslant z$ imply that

$$w \vee y \leqslant x \vee z \quad \text{and} \quad w \wedge y \leqslant x \wedge z.$$

1.13. Prove that in any lattice L,

$$x \wedge (y \vee z) \geqslant (x \wedge y) \vee (x \wedge z),$$
$$x \vee (y \wedge z) \leqslant (x \vee y) \wedge (x \vee z),$$

for all $x, y, z \in L$. (These formulae are called the *one-sided distributive laws*.)

1.14. Let L be a distributive lattice, and let L' be the lattice obtained by adding to L two elements a and b such that $a < x < b$ for all $x \in L$. Prove that the lattice L' is distributive.

1.15. A lattice L is said to be *modular* if $x \leqslant z$ implies that $x \vee (y \wedge z) = (x \vee y) \wedge z$, for all $x, y, z \in L$. Prove that every distributive lattice is modular.

1.16. Prove that for any elements x and y of a Boolean algebra,

$$x \vee (\bar{x} \wedge y) = x \vee y \quad \text{and} \quad x \wedge (\bar{x} \vee y) = x \wedge y.$$

2 Graphs and algorithms

2.1. Introduction

IN THIS chapter we shall first present the basic concepts of graph theory, and then consider some particular kinds of graphs which are of special importance from both a theoretical and a practical viewpoint. In the course of our discussion we shall develop a number of methods of analysing graphs; these will usually be presented in the form of algorithms, from which computer programs can easily be constructed.

Many graph-theoretic algorithms are deceptively simple in appearance: in practice they may require a prohibitive amount of work. The final section of this chapter introduces the notion of the complexity of an algorithm—which is a measure of the work involved in executing it—and explains how this complexity can be determined.

2.2. Graphs

2.2.1. Definition of a graph

A *graph* $G = (X, U)$ consists of

(i) a finite set $X = \{x_1, x_2, \ldots, x_n\}$, whose elements are called *nodes*, and
(ii) a subset U of the Cartesian product $X \times X$, the elements of which are called *arcs*.

A graph can be depicted by a diagram in which nodes are represented by points in the plane, and each arc (x_i, x_j) is indicated by an arrow drawn from the point representing x_i to the point representing x_j. For example, Fig. 2.1 represents the graph $G = (X, U)$ where

$$X = \{x_1, x_2, x_3, x_4\},$$

$$U = \{(x_1, x_2), (x_1, x_4), (x_2, x_2), (x_2, x_4), (x_3, x_2), (x_4, x_1), (x_4, x_3)\}.$$

In discussing graphs we shall use the following terminology:

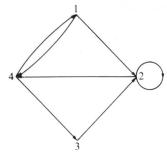

FIG. 2.1

Initial and terminal endpoints of an arc. For an arc (x_i, x_j), the node x_i is the *initial endpoint* and the node x_j is the *terminal endpoint*. An arc whose endpoints are coincident, i.e. an arc of the form (x_i, x_i), is called a *loop*.

Adjacent nodes, adjacent arcs. Two nodes are said to be *adjacent* if they are joined by an arc. Two arcs are *adjacent* if they have at least one common endpoint.

Successors, predecessors, and neighbours of a node. In a graph $G = (X, U)$, a node x_j is called a *successor* of a node x_i if $(x_i, x_j) \in U$; the set of all successors of x_i is denoted by $\Gamma^+(x_i)$. Similarly, a node x_j is called a *predecessor* of x_i if $(x_j, x_i) \in U$, and the set of all predecessors of x_i is denoted by $\Gamma^-(x_i)$. A node which is either a predecessor or a successor of a node x_i is sometimes called a *neighbour* of x_i; the set of all neighbours of x_i is denoted by $\Gamma(x_i)$. It is evident that $\Gamma(x_i) = \Gamma^+(x_i) \cup \Gamma^-(x_i)$.

Arcs incident to and from a node. If an arc u has node x_i as its initial endpoint, we say that the arc u is *incident from* x_i; whereas if an arc u has node x_i as its terminal endpoint we say that arc u is *incident to* x_i. The number of arcs incident from a node x_i is called the *exterior semi-degree* or the *out-degree* of x_i, and it is denoted by $\rho^+(x_i)$; while the number of arcs incident to x_i is called the *interior semi-degree* or *in-degree* of x_i, and it is denoted by $\rho^-(x_i)$.

Example 2.1. In the graph of Fig. 2.1, the node x_2 has the set of successors $\Gamma^+(x_2) = \{x_2, x_4\}$ and the set of predecessors $\Gamma^-(x_2) = \{x_1, x_2, x_3\}$;

the out-degree of this node is $\rho^+(x_2) = |\Gamma^+(x_2)| = 2$, and its in-degree is $\rho^-(x_2) = |\Gamma^-(x_2)| = 3$.

Partial graphs. If we remove from a graph $G = (X, U)$ a subset of its arcs, we are left with a graph of the form

$$H = (X, V) \quad \text{where} \quad V \subseteq U,$$

which is called a *partial graph* of G. The graph of Fig. 2.3 is a partial graph of that shown in Fig. 2.2.

Subgraphs. If we remove from a graph $G = (X, U)$ a subset of its nodes, together with all the arcs incident to or from those nodes, we are left with a graph of the form

$$H = (Y, U_Y) \quad \text{where} \quad Y \subseteq X \quad \text{and} \quad U_Y = U \cap (Y \times Y),$$

which is called a *subgraph* of G. We may describe H more precisely, as the subgraph of G *generated by* Y. As an example, Fig. 2.4 shows a subgraph of the graph in Fig. 2.2, in particular the subgraph generated by $\{x_1, x_2, x_3\}$.

Condensations. Let $G = (X, U)$ be any graph, and let $\mathscr{S} = \{X_1, X_2, \ldots, X_n\}$ be a partition of the node-set X of G. Then the *condensation of G induced by $\mathscr{S}$* is defined as the graph $G_{\mathscr{S}} = (\mathscr{S}, U_{\mathscr{S}})$, where

$$U_{\mathscr{S}} = \{(X_r, X_s) \in \mathscr{S} \times \mathscr{S} | X_r \neq X_s, \text{ and for some } x_i \in X_r$$

$$\text{and some } x_j \in X_s, (x_i, x_j) \in U\}.$$

In pictorial terms, $G_{\mathscr{S}}$ is obtained from G by coalescing the nodes of each member of $\mathscr{S}$, and then removing any loops. The condensation

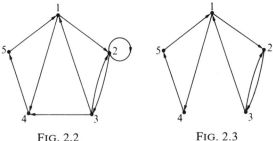

FIG. 2.2   FIG. 2.3

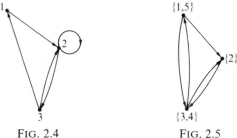

FIG. 2.4   FIG. 2.5

of the graph of Fig. 2.2 which is induced by the partition $\{\{x_1, x_5\}, \{x_2\}, \{x_3, x_4\}\}$ is shown in Fig. 2.5.

2.2.2. *Graphs and relations*

Each graph $G = (X, U)$ determines a binary relation $\mathcal{R}_G$ on the set X of its nodes, through the rule

$$x_i \mathcal{R}_G x_j \quad \text{if and only if} \quad (x_i, x_j) \in U.$$

(Using the terminology of the previous section, $\mathcal{R}_G$ is the relation 'is a predecessor of' on the node-set X.)

Conversely, each binary relation $\mathcal{R}$ on a set X determines a graph $G = (X, U_{\mathcal{R}})$, where

$$U_{\mathcal{R}} = \{(x_i, x_j) \in X \times X \mid x_i \mathcal{R} x_j\},$$

which is called the *graph of* $\mathcal{R}$. As an illustration, Fig. 2.7 shows the graph of the relation 'was a parent of' on a set of nine Greek gods, whose genealogical tree is given in Fig. 2.6.

As one might therefore expect, the concepts introduced in our discussion of relations in Section 1.4 all have counterparts in the theory of graphs. Indeed, just as a relation $\mathcal{R}$ on a set has a complement $\bar{\mathcal{R}}$ and a converse $\mathcal{R}'$ (see Section 1.4.2), a graph

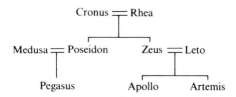

FIG. 2.6

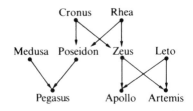

FIG. 2.7

$G = (X, U)$ has a complement $\bar{G} = (X, \bar{U})$ where

$$\bar{U} = \{(x_i, x_j) \in X \times X \mid (x_i, x_j) \notin U\}$$

and a converse $G' = (X, U')$ where

$$U' = \{(x_i, x_j) \in X \times X \mid (x_j, x_i) \in U\}.$$

The complement and the converse of the graph of Fig. 2.1 are shown in Figs. 2.8 and 2.9 respectively.

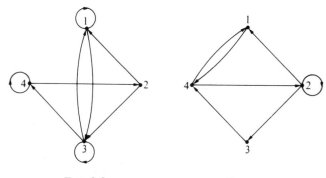

FIG. 2.8   FIG. 2.9

Again, in accordance with our nomenclature for special kinds of relations (see Sections 1.4.3, 1.4.5), we say that a graph $G = (X, U)$ is:

Reflexive	when $(x_i, x_i) \in U$	for all $x_i \in X$;
Anti-reflexive	when $(x_i, x_i) \notin U$	for all $x_i \in X$;
Symmetric	when $(x_i, x_j) \in U$	implies $(x_j, x_i) \in U$;
Anti-symmetric	when $(x_i, x_j) \in U$	and $(x_j, x_i) \in U$ together imply $x_i = x_j$;

Transitive	when $(x_i, x_j) \in U$ and $(x_j, x_k) \in U$ together imply $(x_i, x_k) \in U$;
Complete	when $(x_i, x_j) \notin U$ and $(x_j, x_i) \notin U$ together imply $x_i = x_j$.

Example 2.2. The graph of Fig. 2.10 is reflexive, anti-symmetric, transitive, and complete. Accordingly, the relation 'is a predecessor of' is a total ordering of its node set (as defined in Section 1.4.5).

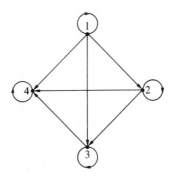

FIG. 2.10

2.2.3. Simple graphs

A graph is said to be *simple* if it is anti-reflexive and symmetric. As an example, the graph of Fig. 2.11(a) is simple.

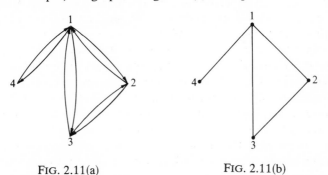

FIG. 2.11(a)   FIG. 2.11(b)

In a simple graph the arcs occur in pairs, the two arcs in each pair having the same endpoints but opposite orientations. In analysing such a graph it is often convenient to consider each arc pair as a single entity; in this case we describe a pair of arcs (x_i, x_j) and (x_j, x_i)

as an *edge* and denote it by $[x_i, x_j]$, or by $[x_j, x_i]$. (We observe that an edge of a graph is essentially a two-element subset of its node set; the order in which these elements are listed has no significance.) The set of edges of a simple graph is usually denoted by E, and we may write $G = (X, E)$.

To represent a simple graph pictorially we often draw each edge $[x_i, x_j]$ as a single undirected line, rather than a pair,of arrows, between the points representing x_i and x_j; as an example, Fig. 2.11(b) shows this type of representation of the graph of Fig. 2.11(a). Diagrams of this kind are frequently encountered, for instance as street maps or wiring diagrams of electric circuits.

For a simple graph, the terms 'predecessor', 'successor', and 'neighbour' are all synonymous. Also, for each node x_i, the in-degree $\rho^-(x_i)$ is equal to the out-degree $\rho^+(x_i)$; we may describe this number simply as the *degree* $\rho(x_i)$ of x_i. When an edge e has a node x_i as one of its endpoints we say that the edge e is *incident with* x_i. Obviously, the degree $\rho(x_i)$ of a node x_i is equal to the number of edges incident with x_i.

We remark that, although graphs are often presented as being of two different kinds—namely 'directed' graphs and 'undirected' graphs—here we consider all graphs to be 'directed', and we regard an 'undirected' graph as a 'directed' graph which is anti-reflexive and symmetric. Of course, in constructing a graph model of a physical system such as a traffic network or an electric circuit, it may be more natural to relate the components of the system (such as streets or electrical conductors) to edges than to arcs, and the subsequent analysis of the graph may be easier to perform in terms of edges. However, there are many graph concepts and analysis methods which are relevant to all graphs, simple and otherwise, and which we shall want to present in their most general form. In order to apply these methods to a simple graph, it will only be necessary to decompose its edges into their constituent arcs. This decomposition of edges into arcs usually has a simple physical interpretation: we note for instance that a two-way street is essentially a pair of contiguous one-way streets which carry traffic in opposite directions.

The simplification of a graph. Any graph $G = (X, U)$ determines a simple graph $G_s = (X, U_s)$, where $U_s = \{(x_i, x_j) \in X \times X \mid x_i \neq x_j$ and either $(x_i, x_j) \in U$ or $(x_j, x_i) \in U$}. This graph G_s is called the *simple graph associated with G*, or the *simplification of G*. As an example, the simplification of the graph of Fig. 2.12(a) is shown in Fig. 2.12(b).

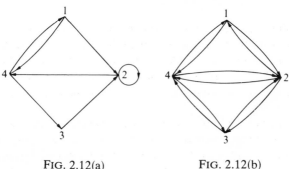

FIG. 2.12(a)   FIG. 2.12(b)

2.2.4. p-graphs and multigraphs

The concept of a graph can be extended by allowing a pair of nodes to be joined by several distinct arcs with the same orientations (see Fig. 2.13). A schema of this kind is called a *p-graph*, where *p* is the maximum number of arcs having the same initial endpoints and the same terminal endpoints. (Thus, the schema of Fig. 2.13 is a 3-graph.)

The notion of a simple graph can also be extended, by allowing a pair of nodes to be joined by more than one edge (see Fig. 2.14); a schema of this kind is called a *multigraph*. We shall make some use of these concepts in later chapters.

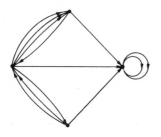 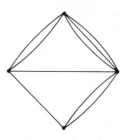

FIG. 2.13   FIG. 2.14

2.3. Paths on graphs

2.3.1. Paths and cycles

A *path* is a finite sequence of arcs of the form

$$\mu = (x_{i_0}, x_{i_1}), (x_{i_1}, x_{i_2}), \ldots, (x_{i_{r-1}}, x_{i_r}),$$

i.e. a finite sequence of arcs in which the terminal node of each arc coincides with the initial node of the following arc; the number r of arcs in the sequence is called the *order* of the path. The initial endpoint of the first arc and the terminal endpoint of the last arc of a path are called respectively the *initial* and *terminal endpoints* of the path.

A path whose endpoints are distinct is said to be *open*; whereas a path whose endpoints coincide is called a *closed path*, or *cycle*.

A path is *simple* if it does not traverse any arc of a graph more than once. A path is *elementary* if it does not traverse any node more than once, i.e. if all the initial endpoints (or all the terminal endpoints) of its arcs are distinct. It is evident that every elementary path is simple.

It will be observed that a path is completely determined by the sequence of nodes $x_{i_0}, x_{i_1}, x_{i_2}, \ldots, x_{i_r}$ which it visits; we shall often find it convenient to specify a path by listing this node sequence rather than the arc sequence.

Example 2.3. In the graph of Fig. 2.1, the arc sequence

$$(x_1, x_4), (x_4, x_3), (x_3, x_2), (x_2, x_4), (x_4, x_1), (x_1, x_2)$$

is a path of order 6, from x_1 to x_2; this path is simple, but it is not elementary. The same path is described by the node sequence $x_1, x_4, x_3, x_2, x_4, x_1, x_2$. The arc sequence

$$(x_1, x_4), (x_4, x_3), (x_3, x_2)$$

is an elementary path from x_1 to x_2, and the arc sequence

$$(x_2, x_4), (x_4, x_1), (x_1, x_2)$$

is an elementary cycle.

Example 2.4. *Finite-state systems.* Many problems involve a 'system' which at any time can be in only one of a finite number of different 'states'. A system of this kind can be represented by a *state diagram*, a graph whose nodes correspond to the system states and whose arcs represent the possible direct transitions from one state to another. This form of representation is helpful in investigating the possible 'modes of behaviour' of the system, since the paths on the state diagram determine the possible sequences of state transitions.

Before considering more serious applications we shall illustrate these ideas with a well-known puzzle: A ferryman (f) has to take a wolf (w), a goat (g), and a cabbage (c) across a river. His skiff is so small that, besides the ferryman, it can take only one of the other objects. The wolf cannot be left alone with the goat, nor the goat with the cabbage. How must the ferryman proceed?

We may consider that the ferryman, wolf, goat, and cabbage form a system, whose state at any time can be described by listing the objects left on the first bank of the river. The state diagram, showing all possible direct transitions, is given in Fig. 2.15. (The graph is symmetric because, for this particular system, all transitions are reversible.) The solutions to the problem are given by the paths from the 'initial state' node to the 'final state' node.

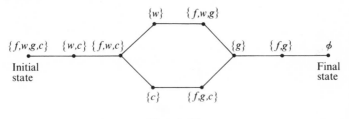

FIG. 2.15

As another example, an important concept in computer science is that of a *finite-state machine*, which consists of

(1) a finite set S of *states*,
(2) a finite set X of *input symbols*,
(3) a finite set Y of *output symbols*,
(4) a *transition function* $f: S \times X \to S$,
(5) an *output function* $g: S \times X \to Y$.

A finite-state machine acts by 'reading' a string of input symbols and 'writing' a string of output symbols in the following manner. If the machine is currently in the state $s \in S$ and it is presented with an input symbol $x \in X$, then it will change its state to $f(s, x)$ and write the output symbol $g(s, x)$.

To illustrate the manner in which such machines operate, Fig. 2.16 shows the state diagram of a machine which recognizes every sequence '101' in a string of zeros and ones. In this diagram, the first label on each arc is the input symbol which causes the corresponding transition; the second label is the symbol which is written when this transition takes place. The response of this machine to a particular input string is shown below, for the case where the machine is initially in the state 'a'.

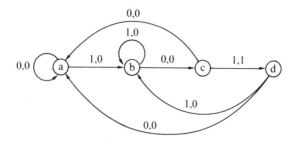

FIG. 2.16

Input string:	0 1 0 0 1 0 1 1 1 0 1 1 0
Successive states:	a | a | b | c | a | b | c | d | b | b | c | d | b | c
Output string:	0 0 0 0 0 0 1 0 0 0 1 0 0

Finite-state machines can easily be simulated by computer programs, and programmed models of 'recognizers' of the type described above are used for lexical analysis in compilers, to recognize program identifiers and 'reserved' words such as *if*, *do*, *go to*, etc. (Aho and Ullman 1977). Finite-state machine concepts are also used in designing hardware for digital systems (Clare 1973).

2.3.2. Chains on simple graphs

In a simple graph $G = (X, E)$, a *chain* is defined to be a finite sequence of edges, of the form

$$\sigma = [x_{i_0}, x_{i_1}], [x_{i_1}, x_{i_2}], \ldots, [x_{i_{r-1}}, x_{i_r}],$$

i.e. a sequence of edges in which each edge has one endpoint in common with the preceding edge, and the other endpoint in common with the following edge. The nodes x_{i_0} and x_{i_r} are called the *endpoints* of the chain. A chain is *open* if its endpoints are distinct; it is *closed* if the endpoints are coincident.

A chain is *simple* if it does not use any edge of a graph more than once. A closed, simple chain is called a *circuit*.

A chain is *elementary* if it does not traverse any node more than once. Every elementary chain is simple.

Example 2.5. In the graph of Fig. 2.11, the edge sequence $[x_1, x_2]$, $[x_2, x_3], [x_3, x_1], [x_1, x_4]$ is a simple but non-elementary chain, from x_1 to x_4. The sequence $[x_2, x_3], [x_3, x_1], [x_1, x_4]$ is an elementary chain from x_2 to x_4, and the sequence $[x_1, x_2], [x_2, x_3], [x_3, x_1]$ is a circuit.

2.4. Some forms of connectedness of graphs

2.4.1. Accessibility

Let $G = (X, U)$ be any graph, and let x_i be one of its nodes. Then any node x_j (not necessarily distinct from x_i) such that there exists a path from x_i to x_j is called a *descendant* of x_i; while any node x_j (not necessarily distinct from x_i) such that there exists a path from x_j to x_i is called an *ascendant* of x_i. It will be observed that a node x_j can be both a descendant and an ascendant of x_i: this occurs whenever there exists a cycle passing through both x_i and x_j. We shall denote the set of descendants of a node x_i by $\hat{\Gamma}^+(x_i)$, and the set of its ascendants by $\hat{\Gamma}^-(x_i)$.

A node x_j is said to be *accessible* from a node x_i if x_j is a descendant of x_i or $x_j = x_i$; similarly, x_j is said to be *converse-accessible* from x_i if x_j is an ascendant of x_i or $x_j = x_i$. The sets of nodes which are accessible and converse-accessible from x_i will be denoted by $\overset{*}{\hat{\Gamma}}{}^+(x_i)$ and $\overset{*}{\hat{\Gamma}}{}^-(x_i)$ respectively. Clearly,

$$\overset{*}{\hat{\Gamma}}{}^+(x_i) = \{x_i\} \cup \overset{*}{\hat{\Gamma}}{}^+(x_i), \quad \text{and} \quad \overset{*}{\hat{\Gamma}}{}^-(x_i) = \{x_i\} \cup \overset{*}{\hat{\Gamma}}{}^-(x_i).$$

Example 2.6. On the graph of Fig. 2.7,

$\hat{\Gamma}^+(\text{Cronus}) = \{\text{Poseidon, Zeus, Pegasus, Apollo, Artemis}\}$,
$\hat{\Gamma}^-(\text{Pegasus}) = \{\text{Poseidon, Cronus, Rhea, Medusa}\}$.

Example 2.7. On the graph $G = (X, U)$ of Fig. 2.1, $\hat{\Gamma}^+(x_i) = X$ and $\hat{\Gamma}^-(x_i) = X$, for all $x_i \in X$.

2.4.2. Connectivity

For any graph $G = (X, U)$, we define a binary relation called the *connectivity relation* C on the node set X by the rule:

$x_i C x_j$ if x_j is accessible from x_i on the simplification G_s of G. When $x_i C x_j$, we say that x_i is *connected to* x_j.

Now the relation C is obviously reflexive, symmetric and transitive, and since it has these properties, it is an equivalence relation.

Consequently, it induces a partition X/C of the node-set of G, whose members are the equivalence classes with respect to C (see Section 1.4.4). The subgraphs of G which are generated by these equivalence classes are called the *connected components* of G.

If a graph has only one connected component, we say that the graph is *connected*.

Example 2.8. The graph $G = (X, U)$ of Fig. 2.17 has two connected components; these are the subgraphs of G generated by $\{x_1, x_2, x_3, x_6, x_7, x_8\}$ and $\{x_4, x_5\}$.

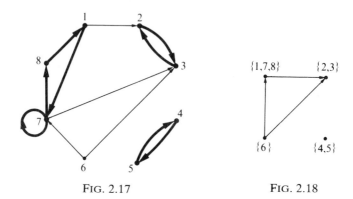

FIG. 2.17   FIG. 2.18

2.4.3. Strong connectivity

Again, let $G = (X, U)$ be any graph, and let us define another binary relation S on the node-set X of G, by the rule:

$x_i \mathsf{S} x_j$ if on G, each of the nodes x_i and x_j is accessible from the other. S is called the *strong connectivity relation* on G, and when $x_i \mathsf{S} x_j$ we say that x_i is *strongly connected* to x_j.

It is easily verified that like C, the relation S is an equivalence relation. The subgraphs of G which are generated by the equivalence classes with respect to S are called the *strongly connec-ted components* of G. If G contains only one strongly connected component, we say that *G is strongly connected*. Clearly, every strongly connected graph is connected.

Example 2.9. The graph $G = (X, U)$ of Fig. 2.17 has four strong components, namely the subgraphs of G generated by $\{x_1, x_7, x_8\}$, $\{x_2, x_3\}$, $\{x_4, x_5\}$ and $\{x_6\}$. The arcs which belong to these components are indicated by bold lines.

In studying the structural properties of a graph $G = (X, U)$, it is often helpful to determine the set of equivalence classes X/S, and then to construct the condensation of G induced by X/S (as defined in Section 2.2). This particular condensation, which is called the *reduced graph* G^R of G, shows clearly the form of connectivity between each pair of nodes on G. (See for instance Fig. 2.18 which shows the reduced form of the graph of Fig. 2.17.) In particular, it is easily verified that if x_i and x_j are nodes of G, which belong respectively to members X_a and X_b of X/S, then

(i) x_i is connected to x_j on G if and only if X_a is connected to X_b on G^R, and

(ii) x_j is accessible from x_i on G if and only if X_b is accessible from X_a on G^R.

Finally, it is important to note that for any graph G, the reduced graph G^R does not contain any cycles. This can be proved by contradiction: Suppose G^R contains a cycle, and let (X_a, X_b) be one of its arcs. Then on G^R, X_a is strongly connected to X_b, which implies that in G there exist nodes $x_i \in X_a$ and $x_j \in X_b$ which are strongly connected. Now the nodes x_i and x_j belong to the same member of X/S, and so $X_a = X_b$. But then, by the method of construction of G^R (see Section 2.2), this graph cannot contain the arc (X_a, X_b), which contradicts our assumption that G^R contains a cycle.

Some special properties of graphs without cycles will be presented in Section 2.6.

2.4.4. *Algorithms for finding accessible sets and the components of a graph*

Accessible sets. The accessible set of a specified node x_k can be found by a simple 'labelling' algorithm, which is described in pictorial terms in Fig. 2.19(a). This algorithm progressively labels the node x_k and its descendants with the symbol $*$, in such a way that on termination all the members of $\overset{*}{\Gamma}{}^+(x_k)$ have received this label. In the course of the algorithm each member x_i of $\overset{*}{\Gamma}{}^+(x_k)$ also bears a label $\dagger$, from the time when x_i is labelled with $*$ until its set of successors $\Gamma^+(x_i)$ is explored.

To implement this algorithm on a computer we must first choose *data structures*, i.e. methods of arranging items of data, which

Step 1	Label node x_k with the symbols † and *.
Step 2	Let x_i be any node labelled with the symbol †. Delete the label † of x_i and label with both † and * every successor of x_i which does not have the label *.
Step 3	If the graph still contains a node labelled with † then return to *Step 2*.
End	The nodes with the label * form the set $\overset{*}{\Gamma}{}^{+}(x_k)$.

FIG. 2.19(a)

enable us to represent the graph and the assignment of labels to its nodes conveniently in the computer memory. For instance, we may choose to define the graph by specifying, for each node x_i, a *list* L_i of the indices of its successors. The set of nodes labelled † can also be recorded conveniently in a list T. To define the set of nodes labelled with * we could again use a list, but we shall find it more appropriate to represent this set by its *characteristic vector*, i.e. a vector $\mathbf{c} = [c(1), c(2), \ldots, c(n)]$ of n elements (where n is the number of nodes in G) in which $c(i) = 1$ if node x_i is labelled with * and $c(i) = 0$ otherwise.

A more precise version of the algorithm of Fig. 2.19(a), using these choices of data structures, is given in Fig. 2.19(b). In this description, the symbolism '$p \leftarrow q$' is read as 'Assign to the variable p the value of q'. This version of the algorithm prints out the indices of the nodes of $\overset{*}{\Gamma}{}^{+}(x_k)$—see Step 2 and Step 6.

With regard to the choice of data structures, we could of course have represented the set of elements labelled with † by its characteristic vector, rather than by the list T, but this would involve more

Step 1	[Initialize] Clear the list T, and set $c(i) \leftarrow 0$ for $i = 1, 2, \ldots, n$.
Step 2	[Label x_k] Enter the index k in the list T, set $c(k) \leftarrow 1$ and print the index k.
Step 3	[Find some node x_i labelled with †] Let i be any index in T. Delete i from T.
Step 4	[Is list of successors of x_i empty?] If L_i is empty then go to *Step 7*.
Step 5	[Choose a successor x_j of x_i] Let j be any index in L_i. Delete j from L_i.
Step 6	[If x_j is not labelled with * then label it with † and *] If $c(j) = 0$ then enter the index j in T, set $c(j) \leftarrow 1$, and print the index j. Return to *Step 4*.
Step 7	[Does graph still contain any nodes with label †?] If T is not empty then return to *Step 3*.
End	

FIG. 2.19(b)

work. (To find a node labelled with †, in Step 3, we would have to scan the elements of this characteristic vector one by one, searching for an element with the value '1'; in the worst case this could involve the inspection of *all* the elements of the vector.) To represent the set of nodes labelled with ∗, we chose to use the characteristic vector **c** rather than a list because in Step 6 we wish to test whether a *particular node x_j* has the label ∗; with the vector representation this only involves the examination of one element, $c(j)$ of the vector, whereas in a list representation we might in the worst case have to scan *all* the entries in the list. (The choice of data structures for graph· algorithms is discussed in detail by Aho, Hopcroft, and Ullman, 1974.)

To find the converse-accessible set of a specified node, the same algorithm can be used: for this purpose we simply specify in each list L_i the predecessors, rather than the successors, of x_i.

Determination of the connected components of a graph. From the definition of connectivity in Section 2.4.2 it follows immediately that for any node x_k, the set of nodes which are connected to x_k on G is identical to the accessible set of x_k on the simplification G_s of G. Hence the algorithm of Fig. 2.19(b) will list the nodes of the connected component which contains any specified node x_k of G, if each list L_i comprises the indices of the *neighbours* of x_i on G.

To find all the components of a graph, we can proceed as follows. First we arbitrarily choose some node x_k of G, and find the node set of the component containing x_k by the labelling method described above. If all the nodes of G are labelled by this process, the problem is solved. Otherwise we arbitrarily select some node which has not yet been labelled and find its component, again by labelling. By repeating this process until all the nodes are labelled, we obtain successively the node sets of all the connected components of G. A detailed version of this algorithm is presented in Fig. 2.20.

Determination of strong components. From the definition of strong connectivity it follows immediately that the set $S(x_k)$ of nodes which are strongly connected to a given node x_k can be expressed as

$$S(x_k) = \overset{*}{\Gamma}{}^{+}(x_k) \cap \overset{*}{\Gamma}{}^{-}(x_k).$$

Thus we can obtain the node set of the strong component containing any specified node x_k by constructing the sets $\overset{*}{\Gamma}{}^{+}(x_k)$ and $\overset{*}{\Gamma}{}^{-}(x_k)$

Step 1 [Initialize] Set $k \leftarrow 0$ and set $c(i) \leftarrow 0$ for $i = 1, 2, \ldots, n$.

Step 2 [Increment node index k] Set $k \leftarrow k + 1$.

Step 3 [Has component containing x_k already been labelled?] If $c(k) = 1$ then go to *Step 5*.

Step 4 [Find the nodes of the component containing x_k, by the labelling method.]

    Step 4.1 [Initialize] Clear the list T.

    Step 4.2 [Label x_k] Enter the index k in T, set $c(k) \leftarrow 1$ and print the index k.

    Step 4.3 [Find some node x_i labelled with †] Let i be any index in T. Delete i from T.

    Step 4.4 [Has x_i any successors?] If L_i is empty then go to *Step 4.7*.

    Step 4.5 [Choose a successor x_j of x_i] Let j be any index in L_i. Delete j from L_i.

    Step 4.6 [If x_j is not labelled with * then label it with † and *] If $c(j) = 0$ then enter the index j in T, set $c(j) \leftarrow 1$ and print the index j. Return to *Step 4.4*.

    Step 4.7 [Does graph still contain any nodes with the label †?] If T is not empty then return to *Step 4.3*.

    Step 4.8 [Terminate labelling procedure] Print 'end of component'.

Step 5 [Have all nodes been examined?] If $k \neq n$ then return to *Step 2*.

End

FIG. 2.20

separately, using the labelling algorithm of Fig. 2.19, and then forming the intersection of these sets.

If all the strong components of a graph are to be found, they can be obtained one after the other, by selecting a node of the graph arbitrarily, finding its strong component, then selecting another node whose strong component has not yet been determined, and so on until all nodes have been assigned to components.

Although this is probably the simplest method, it may involve the repeated labelling of some nodes, since a particular node may be accessible, or converse-accessible, from many different components. Alternative methods which are more efficient for graphs containing many components have been devised by Munro (1971) and Tarjan (1972).

2.5. Acyclic graphs

An *acyclic* graph is a graph which does not contain any cycles. In this section, our purpose is to establish some important properties of acyclic graphs in general; some particular kinds of acyclic graphs will be considered in detail in the next section.

First, we shall demonstrate that *any acyclic graph contains at least one node which has no successors, and at least one node which has no predecessors.* The first part is proved by contradiction, as follows: Let $G = (X, U)$ be an acyclic graph, and let us assume that every node on G has at least one successor. Then, starting from any node x_{i_0}, we can find a successor of x_{i_0}, say x_{i_1}, and then a successor of x_{i_1}, say x_{i_2}, and so on, and hence we may construct a path x_{i_0}, x_{i_1}, $x_{i_2}, \ldots, x_{i_r}$ of arbitrarily high order. But since the number of nodes of G is finite, on a path of sufficiently high order some node x_{i_k} say will be encountered twice. Hence x_{i_k} lies on a cycle, which contradicts the assumption that G is acyclic. The second part can be proved by a similar argument (we trace a path 'backwards', rather than 'forwards', from some node x_{i_0}).

This result leads to the important concept of *node rank*, which is defined as follows. Let $G = (X, U)$ be an acyclic graph, and let N_0 be the (non-empty) set of nodes without predecessors on G:

$$N_0 = \{x_i \in X \mid \Gamma^-(x_i) = \phi\}.$$

Now consider the subgraph G_0 of G which is generated by $X - N_0$, i.e. the subgraph obtained by removing from G all nodes in N_0, and all arcs incident from those nodes. If G_0 contains any nodes, then it must contain at least one node without predecessors (since all subgraphs of G are acyclic), and collecting together all such nodes we obtain a non-empty set

$$N_1 = \{x_i \in X - N_0 \mid \Gamma^-(x_i) \subseteq N_0\}.$$

Now consider the subgraph G_1 of G_0 which is generated by $X - (N_0 \cup N_1)$, i.e. the subgraph obtained from G_0 by removing the nodes of N_1, together with all arcs incident from these nodes. Again, if G_1 contains any nodes, then it contains at least one node without predecessors, and the collection of all such nodes constitutes a non-empty set

$$N_2 = \{x_i \in X - (N_0 \cup N_1) \mid \Gamma^-(x_i) \subseteq N_0 \cup N_1\}.$$

Continuing in this manner until all the nodes of G have been removed, we construct successively the (non-empty) sets

$$N_3 = \left\{ x_i \in X - \bigcup_{k=0}^{2} N_k \;\middle|\; \Gamma^-(x_i) \subseteq \bigcup_{k=0}^{2} N_k \right\}$$

$$\cdots \cdots \cdots \cdots \cdots \cdots \cdots \cdots \cdots \cdots$$

$$N_q = \left\{ x_i \in X - \bigcup_{k=0}^{q-1} N_k \;\middle|\; \Gamma^-(x_i) \subseteq \bigcup_{k=0}^{q-1} N_k \right\},$$

where q is the smallest integer such that

$$X - \bigcup_{k=0}^{q} N_k = \phi.$$

From the method of construction of these sets, it is evident that $\{N_1, N_2, \ldots, N_q\}$ is a partition of X.

Let us now assign to each node x_i of G an integer $r(x_i)$, through the rule

$$if \; x_i \in N_k \quad then \; r(x_i) = k.$$

We call $r(x_i)$ the *rank* of node x_i. Clearly, for each node x_i of G

(i) $r(x_i) < r(x_j)$, for all $x_j \in \Gamma^+(x_i)$, and
(ii) if $r(x_i) > 0$, then x_i has at least one predecessor of rank $r(x_i) - 1$.

Example 2.10. In Fig. 2.21 which depicts an acyclic graph, all nodes of the same rank are placed in the same vertical line.

The concept of node rank leads to the following useful characterization of an acyclic graph:

A necessary and sufficient condition for a graph $G = (X, U)$ to be acyclic is that its nodes $x_1, x_2, \ldots, x_n$ can be numbered (i.e. assigned their integer indices) in such a way that if $(x_i, x_j) \in U$ then $i < j$.

To prove that the condition is necessary, we demonstrate that such a numbering scheme does exist for any acyclic graph: Indeed, if

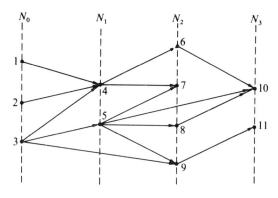

FIG. 2.21

we successively assign the integers, 1, 2, 3, . . . , first to those nodes which are of rank 0, then to the nodes of rank 1, and so on (see, for instance, Fig. 2.21), then we obviously obtain a numbering of the form defined above. The sufficiency of the condition can be demonstrated by contradiction: Let $G = (X, U)$ be a graph which contains a cycle, and let us assume that its nodes are numbered in such a way that if $(x_i, x_j) \in U$ then $i < j$. Now let x_i be the node whose index is maximal on this cycle, and let x_j be its successor on the cycle; then $(x_i, x_j) \in U$ and $i > j$, which contradicts our initial assumption.

Example 2.11. *Critical-path analysis.* A project such as the construction of a large building involves numerous interrelated *activities* (for instance the clearing of a site, laying of foundations, erection of cranes, and so on). In general some activities of a project can take place concurrently, but some cannot begin until certain others have terminated. In organizing a project it is important to know the earliest time at which each activity can begin, and the least time in which the entire project can be completed.

The table below shows how the relationships between the activities of a project may be specified. The activities 1 and 10 are 'dummy' activities of zero duration, representing the project commencement and termination respectively. For each activity i the table gives the activity duration d_i, and a list of *predecessors* of activity i, that is, a list of activities which must all have finished before activity i can commence.

Activity	Duration	Predecessors
1 (start)	0	—
2	4	1
3	10	1
4	6	2
5	2	2
6	11	2
7	22	4, 5
8	3	5
9	17	3, 6, 8
10 (finish)	0	7, 9

From this information one can construct an *activity graph* (Fig. 2.22) to represent the project. In this graph, the nodes correspond to the project activities; more precisely, each node x_i represents an *event*—the *commencement* of activity i. The timing constraints are represented by the arcs of the graph; specifically, if activity i cannot commence before activity k

has finished, that is to say until at least d_k time units after activity k commenced, this condition is represented by an arc (x_k, x_i), which is labelled with d_k. (The label assigned to an arc is called its *length*.) Note that an activity graph is necessarily acyclic, for if a cycle did exist, none of the activities on that cycle could ever commence.

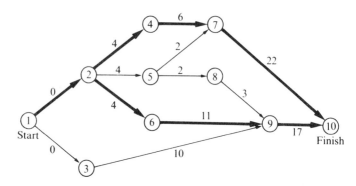

FIG. 2.22

The earliest time t_i at which the ith activity can commence, measured from the start of the project, is obviously given by the length of a longest path from the 'start' node to node x_i. Accordingly, the shortest time in which the entire project can be executed is given by the length of a longest path from the 'start' node to the 'finish' node; for this reason, the longest paths from the 'start' to the 'finish' nodes are called *critical paths*. (The activity graph of Fig. 2.22 has two critical paths, of length 32, which are indicated by bold lines.)

It is evident that activities which do not lie on critical paths can be retarded to some extent without increasing the time needed for the execution of a complete project, and it is useful to have a measure of this latitude: If the earliest possible project completion time is τ, then the *latest* time t'_i at which an activity i can begin without delaying the project completion is given by

$$t'_i = \tau - l_i,$$

where l_i is the length of a longest path from node x_i to the 'finish' node. The maximum amount by which the ith activity can be delayed (its '*slack time*') is then given by the difference

$$s_i = t'_i - t_i.$$

The table below gives the earliest and latest possible starting times, and the slack times, for the project of Fig. 2.22.

Activity:	1	2	3	4	5	6	7	8	9	10
Earliest starting time:	0	0	0	4	4	4	10	6	15	32
Latest starting time:	0	0	5	4	8	4	10	12	15	32
Slack time:	0	0	5	0	4	0	0	6	0	0

A systematic method of obtaining these times, suitable for use on a computer, will be given in the next chapter (Example 3.20). For a more detailed discussion of network models of projects see Elmaghraby (1970) and Roy (1970).

2.6. Trees

2.6.1. Elementary properties of trees

A *tree* is an acyclic graph $G = (X, U)$ in which one node x_r has no predecessors and every other node has exactly one predecessor†. The node x_r is called the *root* of the tree. The graphs shown in Fig. 2.23 are both trees.

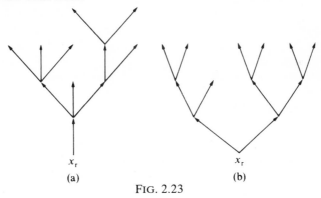

x_r

(a)

x_r

(b)

FIG. 2.23

Trees can be characterized in several different ways, as is demonstrated by the following theorem.

For a graph $G = (X, U)$, the following statements are equivalent:

(1) *G is an acyclic graph in which one node x_r has no predecessors and every other node has exactly one predecessor.*
(2) *G is a connected graph in which one node x_r has no predecessors and every other node has exactly one predecessor.*
(3) *G has a node x_r which is joined to every other node by a unique path from x_r.*

† Some authors describe graphs of this kind as 'arborescences'.

We shall prove this theorem by demonstrating that (1) implies (2), (2) implies (3), and (3) implies (1).

(1) implies (2). In any acyclic graph, each node x_i with rank $r(x_i) \neq 0$ has a predecessor of rank $r(x_i) - 1$, which implies that x_i is accessible from some node of rank zero. Since x_r is the only node of G with rank zero, all the nodes of G are accessible from x_r, which implies that G is connected.

(2) implies (3). Since G is connected, the simple graph G_s associated with G has at least one path from x_r to each of the other nodes. Now let μ be such a path, from x_r to x_i say, and let (x_r, x_k) be the first arc of this path. The existence of (x_r, x_k) in G_s implies that either (x_r, x_k) belongs to G or (x_k, x_r) belongs to G; however, the arc (x_k, x_r) cannot belong to G, since x_r has no predecessors on G, and therefore the arc (x_r, x_k) belongs to G. In the same way, since x_k has only one predecessor on G, the second arc of μ must also belong to G, and by repeating the argument we find that G contains all the arcs of μ. It follows that G contains at least one path from x_r to each of the other nodes. There cannot be more than one path from x_r to any other node x_i, for this would imply that x_i or one of its ascendants had more than one predecessor.

(3) implies (1). G cannot contain a cycle, for otherwise each node on this cycle could be reached from x_r by more than one path. The node x_r has no predecessors, for if x_r had a predecessor x_i then the path from x_r to x_i together with the arc (x_i, x_r) would form a cycle. Each node $x_i \neq x_r$ has at least one predecessor, since x_i is the endpoint of a path from x_r; however, x_i cannot have more than one predecessor, for this would imply the existence of more than one path from x_r to x_i.

Example 2.12. *Lineal charts (family trees).* Let X be a set of people, comprising some individual together with all his or her descendants, and let G be the graph of the relation 'is a parent of' on X. The graph G is acyclic, since no person can be an ancestor of himself; furthermore, if no 'inbreeding' has occurred, G is a tree.

Example 2.13. *Classification systems.* A *classification* of a set X of objects is essentially a partition of X into blocks, which may themselves be partitioned into blocks, and so on. For instance, in the biological classification system the animal and plant kingdoms are classified in this way, the blocks of the successive partitions being known as the *phyla, classes, orders, families, genera,* and *species* of organisms. Any classification system of this kind can be regarded as a tree (see Fig. 2.24).

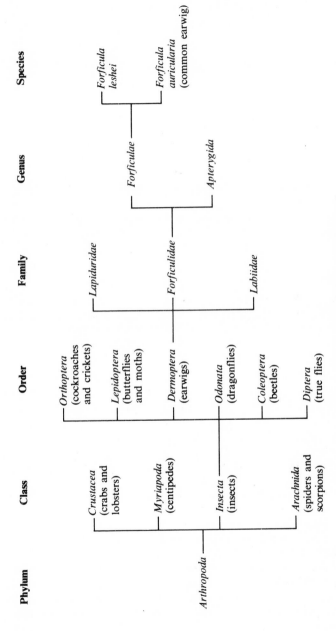

FIG. 2.24. Classification of the common earwig.

The nodes of a tree which have no successors are called its *terminal nodes*.

A tree in which each node has not more than two successors is called a *binary* tree; a binary tree is said to be *complete* if its non-terminal nodes all have exactly two successors. The graph of Fig. 2.23(b) is a complete binary tree.

Example 2.14. *Syntactic trees.* A sentence consists of a number of syntactic entities (such as 'noun phrases' or 'verb phrases') which are concatenated with each other in accordance with certain syntactic or grammatical rules. The process of *parsing* or resolving a sentence into its syntactic components leads naturally to a tree. For example, the syntactic structure of the sentence 'The bird pecked a cherry' is depicted by the tree of Fig. 2.25, whose non-terminal nodes are labelled with syntactic categories, and whose terminal nodes are labelled with the words of the sentence.

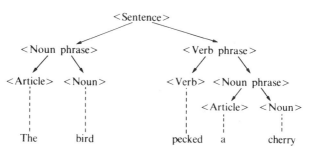

FIG. 2.25

In a similar way, an arithmetic expression such as

$$(a \times (b - c)) + d$$

can be decomposed into a pair of 'sub-expressions' linked by a binary operator symbol (in this instance we have the sub-expressions '$a \times (b - c)$' and 'd', linked by the operator symbol ' $+$ '), and by repeated decomposition of the sub-expressions we again obtain a tree—see Fig. 2.26(a). An alternative way of representing this expression by a tree is shown in Fig. 2.26(b), which is obtained from Fig. 2.26(a) by 'hoisting' each operator symbol to the node above it.

The parsing of a computer program, using the syntactic rules of the programming language, is an important step in program compilation. (Aho and Ullman 1977).

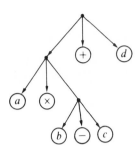

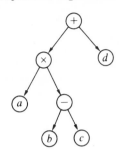

FIG. 2.26(a)   FIG. 2.26(b)

2.6.2. *Transverse orderings of trees*

In an English sentence, or an arithmetic expression, the order of occurrence of the syntactic constituents is usually significant, and in drawing a syntactic tree we take account of this ordering. For instance, when we parsed the sentence 'The bird pecked a cherry' we noted first that the sentence comprised a noun phrase ('The bird') *followed by* a verb phrase ('pecked a cherry'), and in drawing the syntactic tree we indicated this ordering by placing the node representing the verb phrase to the *right* of the node representing the noun phrase (see Fig. 2.25). The same convention was observed in drawing Fig. 2.26, where the relative disposition of the nodes b and c indicates that c is to be subtracted from b, rather than b from c.

In algebraic terms, the node set of each of these trees is equipped with an ordering, of the following kind.

Let $G = (X, U)$ be any tree; then a *transverse ordering* of G is an ordering $\leqslant$ of its node set X such that for any two nodes x_i, x_j,

(i) if $\Gamma^-(x_i) = \Gamma^-(x_j)$ then either $x_i \leqslant x_j$ or $x_j \leqslant x_i$, and
(ii) if $\Gamma^-(x_i) \neq \Gamma^-(x_j)$ then x_i and x_j are incomparable.

Thus, a transverse ordering of a tree provides a total ordering of the successor set of each of its nodes.

Given a transverse ordering $\leqslant$ of a tree $G = (X, U)$, we may construct a total ordering α of X which is compatible with $\leqslant$ (in that for every pair (x_i, x_j) for which $x_i \leqslant x_j$ we also have $x_i \alpha x_j$), in several different ways. As an important example, the *Tarry ordering* α of X is the ordering defined by the rules

(i) if x_i is an ascendant of x_j then $x_i \alpha x_j$, and

(ii) if $x_i \leqslant x_j$ then for all $x_i' \in \overset{*}{\Gamma}{}^+(x_i)$ and all $x_j' \in \overset{*}{\Gamma}{}^+(x_j)$, $x_i' \alpha x_j'$.

Alternatively, if we replace the word 'ascendant' by 'descendant' in rule (i) above, we obtain another total ordering of X called the *reverse Tarry ordering*.

As an example, let us suppose that for the tree of Fig. 2.27 we have the transverse ordering $\leqslant$ where

$$b \leqslant c \leqslant d, \qquad e \leqslant f, \qquad g \leqslant h.$$

(In Fig. 2.27(a) the sets of nodes which are totally ordered by $\leqslant$ are circumscribed by broken lines, and when $x_i \leqslant x_j$ the node x_i is placed to the left of x_j.)

Then in the Tarry order the nodes appear as

$$a \quad b \quad e \quad f \quad c \quad d \quad g \quad i \quad h$$

whereas in the reverse Tarry order we have

$$e \quad f \quad b \quad c \quad i \quad g \quad h \quad d \quad a.$$

For any tree, these orderings can easily be constructed as follows. Let us suppose that, starting from the root, we 'visit' the nodes of the tree in the order defined by the following rule, (cf. Fig. 2.27(b)):

Let x_ν be the node currently being visited. If x_ν has any successors which have not yet been visited then proceed to the least of these (with respect to the transverse ordering $\leqslant$); otherwise return to the predecessor of x_ν (unless x_ν is the root, in which case the procedure is terminated).

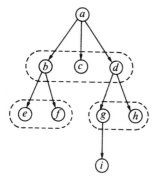

FIG. 2.27(a)

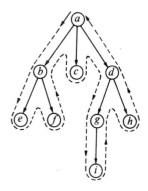

FIG. 2.27(b)

Then the Tarry order (or, respectively, reverse Tarry order) is the order in which the nodes are first (or, respectively, last) visited in this 'traversal' of the tree†.

Example 2.15. *Polish notation.* Let G be a tree representing an arithmetic expression, as in Fig. 2.26(b). Then the listing of the nodes of G in Tarry order is called the *Polish* (or *prefix*) *representation* of the expression, while the listing of the nodes in reverse Tarry order is the *reverse Polish* (or *postfix*) *representation* of the expression. Thus the expression in the usual (*infix*) notation

$$(a \times (b - c)) + d$$

depicted in Fig. 2.26(b) appears in Polish notation as

$$+ \times a - bcd$$

and in reverse Polish notation as

$$abc - \times d + .$$

These notations, introduced by the Polish logician Łukasiewicz, eliminate the need for parentheses or rules of operator precedence to make the syntactic structure of an expression clear. The notations are also of practical importance in connection with program compilation (Aho and Ullman 1977).

Example 2.16. *Succession to the English throne.* Let X be the set comprising a reigning English monarch together with all his or her living descendants, and let G be the graph of the relation '*is a parent of*' on X. It will be assumed that no in-breeding has occurred, so that G is a tree. Now let $\leqslant$ be the transverse ordering of G in which, for any two individuals x_i and x_j which have the same parent in X, $x_i \leqslant x_j$ if either (i) x_i is male and x_j is female or (ii) x_i and x_j are of the same sex and x_i is older than x_j. Then the Tarry order of G is the order of succession to the throne.

Now let us suppose that we are given a complete binary tree $G = (X, U)$, with a transverse ordering $\leqslant$. Then as another example of a total ordering of X which is compatible with $\leqslant$ we have the *symmetric ordering* σ, defined by the following rule: if x_k is any

---

† In computer science these orderings are often called respectively the *pre-order* and *post-order* of the nodes of a tree. However, since the term 'pre-ordering' is commonly used for a relation which is reflexive and transitive (see Section 1.4.3), we prefer the alternative names, after Tarry (1895) who devised a graph traversal algorithm of the kind described here.

non-terminal node of G, having successors x_i and x_j with $x_i \leqslant x_j$, then

$$x_i' \, \sigma \, x_k \quad \text{for every node } x_i' \in \overset{*}{\Gamma}{}^+(x_i),$$

and

$$x_k \, \sigma \, x_j' \quad \text{for every mode } x_j' \in \overset{*}{\Gamma}{}^+(x_j).$$

As an illustration, for the binary tree of Fig. 2.28, with a transverse ordering $\leqslant$ given by

$$b \leqslant c, \quad d \leqslant e, \quad f \leqslant g, \quad h \leqslant i,$$

the listing of the nodes in symmetric order is

$$d \quad b \quad e \quad a \quad f \quad c \quad h \quad g \quad i.$$

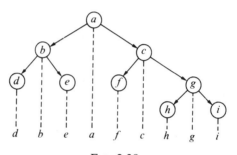

FIG. 2.28

The symmetric ordering of a binary tree can easily be obtained by 'traversing' the tree in the manner described previously and listing the names of nodes as they are first encountered (in the case of terminal nodes) or first revisited (in the case of non-terminal nodes).

Example 2.17. *Binary sort trees.* Let us suppose that we are given a written text, and that we require a listing of all the words which appear in the text, in lexicographic order. One way of solving this problem is to scan the text from the beginning to the end, and in doing so to construct a 'labelled' binary tree, as follows.

Initially, the tree consists of a single node, which is labelled with a 'blank' symbol #. Subsequently, when we read each word ω of the text, we enter the tree at its root, and execute the following algorithm:

Step 1   Let x_ν be the node currently reached, and let l_ν be its label. If l_ν is the symbol '#', go to *Step 2*; if l_ν is a word lexicographically inferior to ω, go to *Step 3*; if l_ν is a word lexicographically superior to ω, go to *Step 4*; and if l_ν is the word ω, go to *End*.

Step 2   Replace the label ♯ of x_ν by the word ω; add two new nodes to the tree—a 'left successor' and a 'right successor' of x_ν—both labelled with ♯; then go to *End*.

Step 3   Advance to the right successor of x_ν, then repeat *Step 1*.

Step 4   Advance to the left successor of x_ν then repeat *Step 1*.

End

As an illustration, Fig. 2.29 shows the tree generated from the following lines of 'The walrus and the carpenter':

> 'The time has come', the Walrus said,
>   'To talk of many things:
> Of shoes—and ships—and sealing-wax—
>   Of cabbages—and kings—

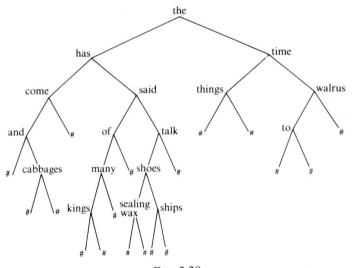

FIG. 2.29

If we now construct the symmetric ordering of the nodes of the tree (by traversing it in the manner indicated previously), and we list the labels of the nodes in this order, we obtain the words of the text in lexicographic order, with a blank symbol before and after each word. Thus for the tree of Fig. 2.29 we obtain

♯ and ♯ cabbages ♯ come ♯ has ♯ kings ♯ . . . ♯ to ♯ walrus ♯.

Binary trees of this kind are often used in compilers, to construct a file (or 'dictionary') giving information on the variables whose names appear in a program text (Knuth 1973).

2.6.3. An algorithm for traversing trees

An algorithm for traversing a binary tree with transverse ordering, in the manner indicated in the previous section, is given in Fig. 2.30. It is assumed that the nodes of the tree are arbitrarily numbered $x_1, x_2, \ldots, x_n$ and that their successors are specified by two vectors **l** and **r** of order n, whose ith entries $l(i)$ and $r(i)$ are the indices of the left and right successors of node x_i; if x_i does not have a left (resp. right) successor then $l(i) = 0$ (resp. $r(i) = 0$).

Step 1   [Initialize] Set $\nu \leftarrow k$ and $\omega \leftarrow k$ (where k is the index of the root node).

Step 2   [Advance to left successor, if this exists] If $l(\nu) \neq 0$ set $\nu \leftarrow l(\nu)$, then set $\omega \leftarrow \omega \circ \nu$, and then repeat *Step 2*.

Step 3   [Advance to right successor, if this exists] If $r(\nu) \neq 0$ set $\nu \leftarrow r(\nu)$, then set $\omega \leftarrow \omega \circ \nu$, and then return to *Step 2*.

Step 4   [Backtrack] Set $\omega \leftarrow sub(\omega)$. If ω is the empty word then go to *End*.

Step 5   [Has right successor been visited?] If $\nu \neq r(last(\omega))$ then set $\nu \leftarrow last(\omega)$ and return to *Step 3*; otherwise set $\nu \leftarrow last(\omega)$ and return to *Step 4*.

End

FIG. 2.30

At each stage of the algorithm, ν is the index of the node which is being visited, while ω is the word (or string) of indices of the nodes which lie on the unique path from the root of the tree to the node x_ν. The symbolism

$$\omega \leftarrow \omega \circ \nu$$

represents the concatenation of the index ν to the word ω (see Example 1.23); also, for any non-empty word ω the notation

$$last(\omega)$$

represents the last symbol in ω, whereas

$$sub(\omega)$$

denotes the word obtained from ω by deleting its last symbol.

It is easy to obtain a Tarry ordering, reverse Tarry ordering or symmetric ordering during the execution of this algorithm: in particular, we can obtain a string σ of the node indices in any one of these orders by inserting the assignment statement

$$\sigma \leftarrow \lambda,$$

where λ is the empty word, in *Step 1*, and inserting the assignment statement

$$\sigma \leftarrow \sigma \circ \nu$$

either at the beginning of *Step 2* (to obtain a Tarry ordering) or in *Step 4* (to obtain a reverse Tarry ordering) or at the beginning of *Step 3* (for a symmetric ordering, in the case where the binary tree is complete).

The application of this algorithm to the tree of Fig. 2.31 is demonstrated in Fig. 2.32, which gives the successive values of ν

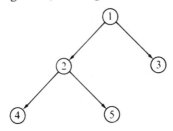

FIG. 2.31

Step	ν	ω	σ_T	σ_{RT}	σ_S
1	1	1	λ	λ	λ
2	2	12	1		
2	4	124	12		
2			124		
3					4
4		12		4	
5	2				
3	5	125			42
2			1245		
3					425
4		12		45	
5	2				
4		1		452	
5	1				
3	3	13			4251
2			12453		
3					42513
4		1		4523	
5	1				
4		λ		45231	

FIG. 2.32

and ω, and also of the strings σ_T, σ_{RT} and σ_S (which eventually represent the Tarry, reverse Tarry and symmetric orders respectively) after the execution of each step.

To implement the algorithm on a computer, the word ω can be represented conveniently by a vector $\mathbf{w} = [w(1), w(2), \ldots, w(n)]$ of n integers, together with an integer variable t (which at each stage gives the *number* of symbols in the word ω). The operations on $\mathbf{w}$ and t which correspond to our operations on the word ω are listed in Fig. 2.33.

Reference or assignment statements		
In terms of word ω	In terms of vector $\mathbf{w}$ and t	In terms of stack S
$\omega \leftarrow \lambda$	$t \leftarrow 0$	clear S
$\omega \leftarrow \omega \circ \nu$	$\begin{cases} t \leftarrow t+1 \\ w(t) \leftarrow \nu \end{cases}$	push ν onto S
$\omega \leftarrow sub(\omega)$	$t \leftarrow t-1$	pop S
$last(\omega)$	$w(t)$	$top(S)$

FIG. 2.33

Alternatively, readers with some knowledge of computer science will perhaps already have visualized these operations in terms of a *stack* (Knuth 1968). This is a store which holds a list or sequence of items of data, with all insertions and deletions being made at one end (the *top* of the stack), in rather the same way as one adds or removes plates from a stack of these; the action of adding an item d of data to a stack is described as *pushing d* onto the stack, whereas the removal of the top item is described as *popping* the stack. (For obvious reasons, a stack is sometimes also called a *last-in–first-out* store or a *push-down* store). The stack operations corresponding to our operations on the word ω are also listed in Fig. 2.33.

2.7. Backtrack programming (or 'tree-search') algorithms

Many combinatorial problems involve the determination of all the elements of a set whose characteristic property is specified. For instance, we may wish to find all the elementary cycles or all the elementary paths between two particular nodes of a given graph. In this section we shall present a systematic method of solving problems of this kind. To introduce the method, it will be convenient to consider first a particular problem.

2.7.1. *The determination of elementary paths*

Let us suppose that for a given graph $G = (X, U)$, we wish to find all the elementary paths from one specified node x_p to another specified node x_q. (It will be assumed here that the nodes x_p and x_q are distinct, and that x_q is accessible from x_p.) We shall denote the set of required paths by M.

Now let $S_p = \{x_{s_1}, x_{s_2}, \ldots, x_{s_k}\}$ be the set of all nodes which are successors of x_p in G, and which are also converse-accessible from x_q in the subgraph H of G obtained by deleting x_p. Also, for $i = 1, 2, \ldots, k$, let M_i denote the subset of paths of M in which the first arc is (x_p, x_{s_i}). Then the set

$$\{M_1, M_2, \ldots, M_k\}$$

is a partition of M. It follows that if we determine the set S_p (which can easily be done, since the required converse-accessible set of x_q can be found by the labelling algorithm of Fig. 2.19), we can 'decompose' the original problem of finding M into k 'sub-problems', involving the separate determination of each of the blocks $M_1, M_2, \ldots, M_k$.

If we now consider the problem of finding one of these blocks M_i, we observe that either

(i) $x_{s_i} = x_q$, in which case M_i contains only one path, which consists of the arc (x_p, x_q), or

(ii) $x_{s_i} \neq x_q$, in which case M_i comprises all the paths obtained by concatenating the arc (x_p, x_{s_i}) with each path in the set M_i' of elementary paths from x_{s_i} to x_q on H.

In case (i) the problem of finding M_i is trivial; in case (ii) the problem reduces to the determination of the set of paths M_i' on H, which is a 'smaller version' of the original problem and which can be decomposed in a similar manner. Thus, by repeated decompositions, we can obtain all the required paths.

As an illustration, let us consider the problem of finding all the elementary paths from x_2 to x_4 on the graph of Fig. 2.34. The successive stages of the decomposition and solution of this problem are depicted by the tree of Fig. 2.35. The root of this tree represents

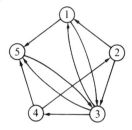

FIG. 2.34

the original problem: inside the root node we have drawn in solid lines the subgraph H of the original graph G which is obtained by removing the node $x_p = x_2$; the asterisks indicate the nodes which belong to the converse-accessible set of the node $x_q = x_4$ on H, and the broken lines indicate the arcs from x_2 which terminate on these nodes. In this particular case $S_2 = \{x_1, x_3\}$, so the original problem can be decomposed into two sub-problems, involving the separate determination of the elementary paths from x_1 to x_4, and from x_3 to x_4, on the subgraph H; these sub-problems are represented by the two tree nodes of rank 1, in which the arcs already assigned to elementary paths are also indicated, by bold lines. The descendants of these tree nodes represent the subsequent decompositions of the sub-problems, and in particular the terminal nodes depict all the trivial sub-problems (and solutions) eventually obtained.

For obvious reasons, the decomposition and solution of a combinatorial problem in this manner is often called a *tree search*.

Implementation of the tree search. With regard to the order in which the sub-problems should be considered after decomposing the original problem we might decompose all the sub-problems which appear as nodes of rank 1 in the search tree, then decompose all the sub-problems of rank 2, and so on; a procedure of this kind is sometimes called a 'breadth-first search'. However, it is evident that if the original graph has many nodes and arcs, the number of sub-problems produced may be very large, in which case a breadth-first search would involve recording a vast amount of data.

As an alternative, we may construct all the sub-problems by repeated application of the following rule:

> *At every stage decompose one of the most recently created sub-problems which remain to be decomposed.*

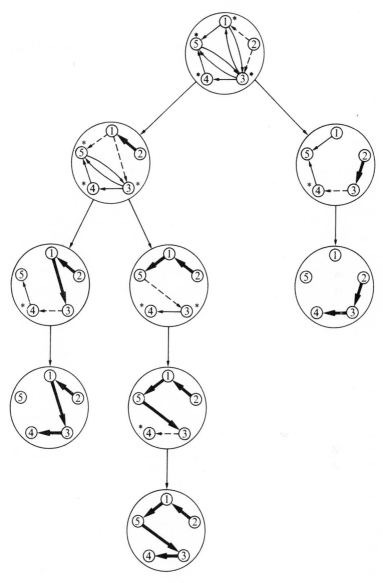

FIG. 2.35

This rule has a simple graph-theoretic interpretation: it states that the problems are to be decomposed in the Tarry order associated with some transverse ordering $\leqslant$ of the search tree. (The transverse ordering $\leqslant$, which corresponds to a total ordering of the set of sub-problems obtained by each decomposition, can be chosen arbitrarily.) Since the decomposition of the sub-problems in a Tarry order effectively involves a traversal of the search tree, this procedure is called a *backtrack programming* or *depth-first search* method. It has the advantage that the only problems which need to be retained at any stage are those which lie on the path from the root of the search tree to the tree node currently being visited, since any other problems which remain to be solved will eventually be derived from problems on this path.

An algorithm for listing elementary paths. A concise description of the above path-finding method is given in Fig. 2.36(a). In this

---

Step 1	[Initialize] $\omega \leftarrow p$, $\nu \leftarrow p$.
Step 2	[Construct list S_ν of successors] Construct S_ν, by the procedure of Fig. 2.36(b).
Step 3	[Have all successors been explored? If not select one] If the list S_ν is empty then go to *Step 5*; otherwise let k be any integer in S_ν, and delete this integer from S_ν.
Step 4	[Destination reached? If so, then print path, otherwise extend path] If $k = q$ then print the word $\omega \circ q$ and return to *Step 3*; otherwise $\omega \leftarrow \omega \circ k$, $\nu \leftarrow k$ and return to *Step 2*.
Step 5 End	[Backtrack] $\omega \leftarrow sub(\omega)$; if $\omega \neq \lambda$ then $\nu \leftarrow last(\omega)$ and return to *Step 3*.

FIG. 2.36(a)

Step *2.1*	[Initialize labelling procedure] Clear the list T. Set $c(i) \leftarrow 0$ and $d(i) \leftarrow 0$ for $i = 1, 2, \ldots, n$. For each index i which appears in ω, set $c(i) \leftarrow 1$.
Step 2.2	[Label x_q] Enter the index q in the list T and set $c(q) \leftarrow 1$ and $d(q) \leftarrow 1$.
Step 2.3	[Label the predecessors of a labelled node] Let r be any index in T. Delete r from T. For each index i which appears in L_r^- and for which $c(i) = 0$, enter index i in T and set $c(i) \leftarrow 1$ and $d(i) \leftarrow 1$.
Step 2.4	[Have all ancestors of x_q been labelled?] If T is not empty return to *Step 2.3*.
Step 2.5	[Form the list S_ν] Clear S_ν. Enter in S_ν every index i which appears in L_ν^+ and for which $d(i) = 1$.

FIG. 2.36(b)

Step	ω	ν	k	S_1	S_2	S_3	S_4	S_5	Output string
1	2	2							
2					13				
3			1		3				
4	21	1							
2				35					
3			3	5					
4	213	3							
2							4		
3			4				ϕ		
4									2134
3									
5	21	1							
3			5	ϕ					
4	215	5							
2								3	
3			3					ϕ	
4	2153	3							
2							4		
3			4				ϕ		
4									21534
3									
5	215	5							
3									
5	21	1							
3									
5	2	2							
3			3			ϕ			
4	23	3							
2							4		
3			4				ϕ		
4									234
3									
5	2	2							
3									
5	λ								

FIG. 2.37

algorithm, the symbols p and q represent the indices of the initial and terminal nodes of the required paths. At each stage, ω is the string of indices of the nodes on the partially formed path from x_p to x_q, and ν is the index of the last of these nodes. (In a programmed version of the algorithm, the string ω can be represented conveniently by a vector and pointer, or by a stack, as indicated in Fig. 2.33.) In *Step 2*, the list S_ν comprises the indices of the nodes of the set

$$\Gamma_G^+(x_\nu) \cap \overset{*}{\Gamma}_H^-(x_q)$$

where $\Gamma_G^+(x_\nu)$ denotes the set of successors of x_ν on the original graph G and $\overset{*}{\Gamma}_H^-(x_q)$ is the converse-accessible set of x_q on the subgraph H of G obtained by removing all the nodes whose indices appear in ω.

The application of this algorithm to the problem of Fig. 2.35 is demonstrated in Fig. 2.37, which gives the values assigned to variables at the execution of each step.

A procedure to compute S_ν (in *Step 2*) is also given, in Fig. 2.36(b). Here the steps 2.1–2.4 construct the characteristic vector **d** of the set $\overset{*}{\Gamma}_H^-(x_q)$, by the labelling method of Section 2.4.4; the final step 2.5 performs the intersection of $\overset{*}{\Gamma}_H^-(x_q)$ with the set $\Gamma_G^+(x_\nu)$, and enters the indices of the nodes of the resulting set in the list S_ν. In this procedure it is assumed that the original graph has been defined by specifying for each node x_i two lists L_i^+ and L_i^-, comprising the indices of the successors and predecessors of x_i respectively.

2.7.2. A general description of backtrack programming

The general concept underlying all backtrack programming methods is that which Polya describes as 'specialization' of a problem (Polya 1957). Let us consider the problem of determining all the elements of some finite set S_0, whose characteristic property is specified: we shall call this 'problem P_0'. To solve the problem we first use some 'specialization rule' to reduce P_0 to a number of sub-problems $P_1, P_2, \ldots, P_k$ where each sub-problem P_i involves the determination of some subset S_i of S_0. The specialization rule is chosen in such a way that each sub-problem P_i is a smaller version of the original problem P_0. Obviously, it is also necessary for every element of S_0 to appear in at least one of the subsets S_i, so that by solving all the sub-problems we obtain every element of S_0. If possible, the sets $S_1, S_2, \ldots, S_k$ should also be pairwise disjoint, for otherwise some element of S_0 will be obtained more than once, and the recognition of repeated elements is troublesome.

After specializing the problem P_0 we specialize each of the sub-problems $P_1, P_2, \ldots, P_k$, and so on, until the original problem

P_0 is reduced to a collection of sub-problems whose solutions are immediately obtainable.

Quite generally, a specialization process of this kind can be regarded as a tree, in which each node represents a problem (or the corresponding subset of S_0 to be determined). More precisely, the root of the tree represents the original problem P_0 (or the required set S_0), and for any node x_i of the tree, the successors of x_i represent the sub-problems (or subsets of S_0) obtained by the specialization of the problem at x_i; while each arc (x_i, x_j) represents the constraint whose addition to the problem at x_i reduces it to the problem at x_j. The terminal nodes represent the trivial sub-problems eventually obtained by the specialization process, whose solutions yield all the required elements of S_0.

To solve a particular problem, the first step is to devise an appropriate specialization rule; a backtrack programming algorithm can then be constructed, by devising a procedure to traverse the search tree determined by the specialization rule.

2.7.3. *The determination of Hamiltonian cycles*

A cycle of a graph is said to be *Hamiltonian* if it traverses every node of the graph exactly once; in the same way, a circuit of a simple graph is said to be Hamiltonian if it traverses every node of the graph exactly once.

To give another demonstration of a tree search, we shall now consider the problem of finding all the Hamiltonian cycles of a given graph $G = (X, U)$. It will be assumed that G has at least two nodes, and no loops.

To present the search method, it will be convenient to view this problem as the problem of finding every subset V of the arc set U of G such that

(i) in the partial graph $H = (X, V)$ of G, every node has exactly one predecessor and exactly one successor, and
(ii) the partial graph $H = (X, V)$ is strongly connected.

We shall denote by M the set of all subsets of U which satisfy these conditions.

Now it is evident that no partial graph of G can be strongly connected unless the graph G is strongly connected. Therefore, as a preliminary step we may construct the accessible sets $\overset{*}{\Gamma}{}^+(x_i)$ and

$\overset{*}{\Gamma}{}^{-}(x_i)$ of some node x_i of G (by the labelling method of Fig. 2.19) and if we find that $\overset{*}{\Gamma}{}^{+}(x_i) \neq X$ or $\overset{*}{\Gamma}{}^{-}(x_i) \neq X$, we know that $M = \phi$ and the search can be terminated.

If we find that G is strongly connected we may next check whether G has exactly two nodes. If so, then $M = \{U\}$ and again the problem is solved. Otherwise, the problem can be decomposed into two 'sub-problems', which are smaller versions of the original one, in the following manner: Let u be any arc of G, and let

$$M_u = \{V \in M \,|\, u \in V\} \quad \text{and} \quad M_{\bar{u}} = \{V \in M \,|\, u \notin V\}.$$

It is evident that

$$M_u \cup M_{\bar{u}} = M \quad \text{and} \quad M_u \cap M_{\bar{u}} = \phi.$$

These two subsets M_u and $M_{\bar{u}}$ of M can be determined separately, as follows.

(i) *The determination of M_u.* Let the initial and terminal endpoints of the arc u be x_i and x_j respectively. Then we define the *contraction of G with respect to u* as the graph G_u obtained from G by (a) removing all the arcs incident from x_i and all the arcs incident to x_j, and also the arc (x_j, x_i) if it exists, and then (b) coalescing the nodes x_i and x_j. As an illustration, Fig. 2.38(b) shows the contraction of the graph of Fig. 2.38(a) with respect to the arc 'a'. It is evident that for each set $V \in M_u$, the arc set $V - \{u\}$ forms a Hamiltonian cycle on G_u; and conversely, for every arc set V' which forms a Hamiltonian cycle on G_u, the arc set $V' \cup \{u\}$ forms a Hamiltonian cycle on G. Thus the problem of finding M_u reduces to the problem of finding the Hamiltonian cycles of G_u.

(ii) *The determination of $M_{\bar{u}}$.* It is evident that the members of $M_{\bar{u}}$ are the Hamiltonian cycles of the partial graph $G_{\bar{u}}$ of G obtained by removing the arc u.

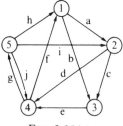

FIG. 2.38(a)

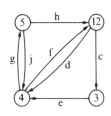

FIG. 2.38(b)

From these arguments it follows that the problem of finding the Hamiltonian cycles of G can be decomposed into two sub-problems, involving the separate determination of the sets of Hamiltonian cycles on the graphs G_u and $G_{\bar{u}}$. These sub-problems may themselves be decomposed, repeatedly, until all the problems become trivial (in that their graphs are either not strongly connected, or have only two nodes).

As an illustration, Fig. 2.39 depicts a search for the Hamiltonian cycles of a 5-node graph (which is drawn inside the root node of the search tree). The label on each tree arc indicates the arc u of G which is used to decompose the problem at each stage. (In this example we have always chosen the arc of G whose name appears first in the alphabetic order, but any other arc could have been chosen.)

It will be observed that, unlike the specialization method employed for finding elementary paths, the specialization method used here can create sub-problems for which the associated subsets of M are null. (For instance, it can be seen on Fig. 2.39 that as a result of two specializations, we create the problem of determining the set

$$\{V \in M \mid a \in V \text{ and } c \notin V\},$$

which is null.) Thus the search tree can have terminal nodes which do not represent Hamiltonian cycles.

The exploration of 'void' problems of this kind can involve a great deal of work, and therefore it is important to eliminate them at an early stage, if possible.

In this connection, it is evident that if a graph G obtained at any stage has any node x_i such that $\rho^+(x_i) = 1$ (resp. $\rho^-(x_i) = 1$), then every Hamiltonian cycle of G must contain the arc u which is incident from x_i (resp. incident to x_i), and therefore $M = M_u$ and $M_{\bar{u}} = \phi$. An arc u of this kind is called an *essential* arc of G. It is profitable, before each specialization of a problem, to search for an essential arc; if one is found, we perform the corresponding contraction immediately, then perform the test for strong connectivity, search for another essential arc, and so on until no further simplifications of this kind can be made. In Fig. 2.39 the broken lines indicate the simplifications obtained by contracting essential arcs.

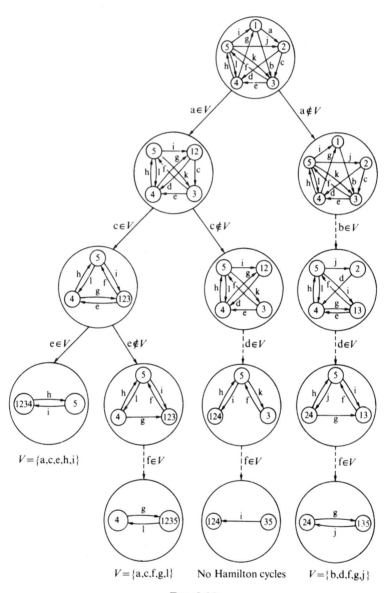

FIG. 2.39

For large graphs, the test for strong connectivity is also useful, in detecting some of the void sub-problems at a relatively early stage. (In fact this is the only reason for performing it, before the number of nodes of a graph has been reduced to two.) Some other ways of eliminating void sub-problems are discussed by Rubin (1974).

Several more examples of backtrack programming will be given in later chapters.

2.8. The time complexity of algorithms

In designing an algorithm, we must obviously ensure that it will be *finite* (i.e. that it will terminate in a finite number of steps) and *correct* (i.e. that it will give the required output and no other output). However, for practical purposes it is also important to consider the amount of work which will be involved in its execution; for instance, if we envisage programming an algorithm for a computer, we would like to have some idea of its running time.

It is obviously very difficult to predict the running time of an algorithm precisely, for it will depend not only on the characteristics of the problem to be solved, but also on the precise way in which the algorithm is programmed, and the instruction execution speeds of the computer which is to be used. However, as we shall demonstrate, it is often possible to estimate the *rate of growth* of the running time, for larger and larger instances of a problem.

To obtain this information, we need to associate with a problem an integer s, called the problem *size*, which is a measure of the amount of input data. For example, the size of a graph-theoretic problem might be the number of nodes of the graph, or the number of its arcs. Then for each step of the algorithm, we establish the *order*† of the time needed to execute it, as a function of s, and we also determine the number of times that the step will be repeated, again as a function of s; with this information we can determine the order of the running time of the complete algorithm, as a function $f(s)$ of the problem size. The function $f(s)$ is described as the *asymptotic time complexity* of the algorithm.

---

† A function $f(s)$ is said to be of order $g(s)$, or in symbols $f(s)$ is $O(g(s))$, if there exists a constant k such that $f(s) \le kg(s)$ for all but some finite (and possibly empty) set of non-negative values of s.

Of course, this function does not define the running time precisely: If we assert that the running time t of an algorithm is $O(s^2)$, we are only saying that this time is expressible in the form

$$t = a_0 + a_1 s + a_2 s^2.$$

However, the order of the running time does indicate the rate of increase of running time with problem size. For instance, if we can establish that the running time of an algorithm is $O(s^2)$, then we know that for large problems, a twofold increase in problem size will increase the running time by a factor of four. Furthermore, this result will hold for any 'straight-forward' programming implementation (without clever tricks or silly inefficiencies), and it will not be affected by changes in computer execution speeds; program details and execution speeds affect the coefficients a_0, a_1, and a_2, but not the *order* of the running time.

Essentially, it is the asymptotic time complexity of an algorithm which determines how large a problem it can solve. To illustrate this point, if a graph-theoretic algorithm has a running time of order 2^n, where n is the number of nodes (and we shall encounter several problems for which the best algorithms available have running times of this order) then an increase of 10 in the number of nodes—say from 10 to 20, or from 20 to 30—will increase the running time by a *factor* of 2^{10}, or approximately 1000. An algorithm whose complexity is of this exponential form† can only solve very small problems, even on the fastest computers. (We observe that for such an algorithm, even a tenfold increase in computer speed only adds three to the size of problem which can be solved in a given time, since $10 \simeq 2^{3.3}$.)

The determination of complexity. Although it is not necessary to consider such details as instruction timing, we cannot establish the complexity of an algorithm until we have specified the *nature* of the machine which is to execute it, and the types of operations which this machine can perform. Here, we shall suppose that our algorithms are to be executed on a 'conventional' digital computer (rather than say, a machine equipped with an associative memory, or some abstract machine capable of unbounded parallelism).

---

† A function $f(s)$ is *exponential* if there exist constants $c_1 > 0$, $k_1 > 1$, $c_2 > 0$ and $k_2 > 1$ such that $c_1 k_1^s \le f(s) \le c_2 k_2^s$ for all but a finite number of values of s.

However, to simplify the analysis it will be convenient to suppose that the random access memory of our computer is unlimited, and that the computer words can store integers of any required size.

To determine the order of the running time of an algorithm on this machine, we will assume that the algorithm is programmed for it in an obvious, straight-forward way. Of course, if an algorithm contains an operation involving objects such as sets or graphs, which can be stored in the memory of a computer in several different ways, it is usually necessary to 'refine' the algorithm by specifying which types of data structures are to be used to define these objects, as the choice of data structure may affect the order of the running time.

Example 2.18. *Complexity of the tree-traversal algorithm.* Let us consider the application of the algorithm of Fig. 2.30, to a complete binary tree which has n nodes. If the word ω is represented by a vector (as indicated in Fig. 2.33), then for each of the steps 1–5 of the algorithm, the time needed for a single execution of the step is independent of n. With regard to the number of times that each step is executed, it is evident that step 1 is executed only once; each of the steps 2–4 are obeyed once for each node of the tree, or n times in all; and step 5 is executed twice for each non-terminal node, that is $n-1$ times in all (see Exercise 2.5). Thus the running time attributable to step 1 is bounded by a constant and the running times attributable to each of the other steps is $O(n)$; hence the total running time of the algorithm is $O(n)$.

With many algorithms, the running time can be substantially different for different problems of the same size. If so, we may wish to know the maximum order of the running time, taken over all problems of a given size; this is called the *worst-case complexity*. Alternatively we may try to determine the 'average' order of the running time over all problems of a given size, which is called the *expected complexity*. The worst-case complexity is usually easier to find than the expected complexity; to obtain the latter we need a realistic probability distribution of the inputs, which may be very hard to specify.

Example 2.19. *Complexity of an algorithm for finding the connected components of a graph.* Let us determine the order of the running time of the algorithm of Fig. 2.20, when applied to a simple graph which has n nodes and e edges. Here, the time taken to execute step 1 is $O(n)$, and it is executed once only; hence the running time attributable to this step is $O(n)$.

The execution times of steps 2, 3, and 5 are each bounded by·a constant, and these steps are performed n times, so the running time attributable to them is also $O(n)$. For the steps 4.1–4.8, the execution times are all bounded by constants; steps 4.1, 4.2, and 4.8 are executed once for each component, and therefore at most n times; steps 4.3 and 4.7 are executed once for each node, or n times in all; step 4.4 is executed $\rho(x_i) + 1$ times for each node x_i of the graph, or $2e + n$ times in all, and steps 4.5 and 4.6 are performed twice for each edge, or $2e$ times in all. Thus, the time attributable to each of the steps 4.1, 4.2, 4.3, 4.7, and 4.8 is at worst $O(n)$, the time attributable to step 4.4 is $O(e + n)$, and the time attributable to steps 4.5 and 4.6 is $O(e)$. The running time of the algorithm is therefore $O(e + n)$.

As a final example we shall establish a bound on the running time of a backtrack programming algorithm. For algorithms of this kind the running times typically increase exponentially with problem size, and this severely limits their usefulness. However, for some problems we do not have any alternative methods, and consequently there is great interest in techniques which attenuate the exponential growth in running time, for instance by 'pruning' branches of a search tree which cannot lead to solutions. Here complexity analyses may be helpful, in providing some indication of the effectiveness of techniques of this kind.

Frequently, the cardinality of the set of elements to be generated by a backtrack algorithm increases very rapidly with problem size (for instance, it is easy to see that an n-node complete and symmetric graph has $(n-1)!$ Hamiltonian cycles.) In such a case it is often most convenient, and informative, to determine the running time in terms of measures of both the input and the output data.

Example 2.20. *Complexity of the backtrack algorithm for listing elementary paths.* For the algorithm of Fig. 2.36(a), it is convenient to establish a bound on the running time in terms of n (the number of nodes of the input graph), a (the number of arcs of this graph), and r (the number of elementary paths from x_p to x_q).

Let us consider first the time taken to execute each step of the algorithm once. For step 1, the execution time is obviously bounded by a constant. The execution time for step 2, as defined in Fig. 2.36(b), can be determined by arguments very similar to those used in the previous example; we leave it as a simple exercise for the reader to verify that this procedure has a running time bound of $O(a + n)$. The execution times of steps 3, 4, and 5 are each bounded by a constant (we ignore the time taken to print the results, in step 4). These execution times are listed in row (i) of Fig. 2.40.

Step	1	2	3	4	5
(i) Maximum time for a single execution	const.	$O(a+n)$	const.	const.	const.
(ii) Maximum number of executions	const.	$O(nr)$	$O(nr)$	$O(nr)$	$O(nr)$
(iii) Maximum running time	const.	$O((a+n)nr)$	$O(nr)$	$O(nr)$	$O(nr)$

FIG. 2.40

With regard to the number of executions of each step, we recall that the algorithm traverses a search tree, whose terminal nodes represent the required elementary paths (see, for instance, Fig. 2.35). The search tree obviously has exactly r terminal nodes and, since the required elementary paths cannot be of order greater than $n - 1$, the tree has at most $(n-2)r+1$ nodes altogether. Now from the definition of the algorithm it is clear that step 1 is executed only once; step 2 is executed once for each non-terminal node of the tree; step 3 is executed once for every node other than the root, with S_ν non-empty, and also once for every non-terminal node, with S_ν empty; step 4 is executed once for every tree node other than the root; and step 5 executed once for every non-terminal node. On this basis, we obtain the bounds on the numbers of executions given in row (ii) of Fig. 2.40.

Combining the results in rows (i) and (ii) of this table we obtain bounds on the running time attributable to each step (these are listed in row (iii)), and hence we find that the total running time of the algorithm has a bound $O((a+n)nr)$. Of course in the worst case, r is itself an exponential function of n.

Easy and hard problems. An algorithm is said to be *polynomial-bounded* if its running time is bounded by a function of order s^k, where s is the problem size and k is a constant. In the context of combinatorial computing, an algorithm is regarded as being *fast* or *efficient* if it is polynomial-bounded, and inefficient otherwise. Accordingly, a problem is said to be *easy* if a polynomial-bounded algorithm has been found for it.

The association of polynomial-boundedness with computational efficiency has a theoretical justification in that, above a certain problem size, a polynomial-bounded algorithm will always have a smaller running time than a non-polynomial-bounded one. Of course, for very small problems the non-polynomial algorithm could have a better performance, but experience indicates that this is not likely to happen in practice.

As we shall see in later chapters, fast algorithms are available for many graph-theoretic problems, but there are a number of important graph problems for which no fast algorithms have ever been found. Among these more difficult problems, we have

(1) *The clique problem*: given a simple graph and an integer k, does the graph have a complete subgraph with k nodes?

(2) *The feedback node set problem*: given a strongly connected graph and an integer k, is it possible to remove k nodes from the graph in such a way as to render it acyclic?

(3) *The feedback arc set problem*: given a strongly connected graph and an integer k, is it possible to remove k arcs from the graph in such a way as to render it acyclic?

(4) *The Hamiltonian cycle problem*: does a given graph contain a Hamiltonian cycle?

(5) *The Hamiltonian circuit problem*: does a given simple graph contain a Hamiltonian circuit?

(6) *The chromatic number problem*: given a graph and an integer k, is it possible to paint the nodes of the graph with k colours, in such a way that no two adjacent nodes are of the same colour?

In principle, all these problems can be solved by tree-search methods, but large instances of the problems are at the moment intractable.

In fact, all the problems listed above belong to a larger class of combinatorial problems (called the 'non-deterministic polynomial-time complete' or *NP-complete* problems), which have been proved to be equivalent, in the sense that either all or none of them can be solved by fast algorithms. (More precisely, it has been shown that each NP-complete problem can be transformed into any other NP-complete problem in polynomial time; clearly, if a problem is easily transformed into an easy problem, then it is also an easy problem, from which it follows that either the NP-complete problems are all easy or none of them are easy.) Since many of these problems have been studied intensively for decades, and no fast algorithms have been found for any of them, it seems likely that no such algorithms exist.

A problem is said to be *hard* if the existence of a fast algorithm for its solution implies that the NP-complete problems are easy. Of course, the NP-complete problems themselves are hard. Some combinatorial optimization problems are also hard, one notorious example being the *Travelling salesman problem*: given a graph in which every arc has a specified length, find a Hamiltonian cycle of minimum length. This problem does not belong to the class of NP-complete problems, but it is obvious that if we had a fast algorithm to solve it, this algorithm could be used to decide quickly whether a given graph has any Hamiltonian cycles at all (see problem (4) above).

Uses and abuses of complexity theory. Quite apart from their intrinsic interest to computer scientists, the techniques and results achieved in complexity analysis are very helpful to those involved in the development of algorithms for practical problems. They enable us to sharpen our otherwise purely intuitive and rather vague notions of the 'efficiency' of algorithms, of 'easy' and 'hard' problems, and they draw attention to the importance of the choice of data structures in implementing graph-theoretic methods. Certainly, in developing any graph algorithm one should establish its worst-case complexity, and when an algorithm cannot be guaranteed to run in less than exponential time one should take a cautious view of its practical feasibility.

At the same time, it is important not to take the worst-case complexity for the expected complexity, and to remember that the worst-case complexity is a measure of the *asymptotic* performance of an algorithm, as the size of its input goes to infinity. There is a tendency to regard an algorithm with a running-time bound of $O(s^2)$ as being 'better' than one with a running-time bound of $O(s^3)$, and yet in practical situations, the latter may invariably have a better performance. Again, in practical applications involving very 'sparse' graphs, the running times of some backtrack algorithms are found to grow as low-order polynomial rather than exponential functions of problem size.

Exercises

2.1. An *Euler cycle* of a graph is a cycle which traverses every arc of the graph once and once only.

   Prove that a connected graph $G = (X, U)$ has an Euler cycle if and only if $\rho^+(x_i) = \rho^-(x_i)$ for all $x_i \in X$.

2.2. Three jealous husbands and their wives want to cross a river. A boat is available, but it is so small that it can only take two people at a time. How can all six people make the crossing, if the husbands will not allow their wives to stay without them in company where other men are present?

2.3. Prove the following:

(i) a graph $G = (X, U)$ is connected if and only if, for any partition $\{X_1, X_2\}$ of X, there exists an arc with one endpoint in X_1 and the other endpoint in X_2, and
(ii) a graph $G = (X, U)$ is strongly connected if and only if, for any partition $\{X_1, X_2\}$ of X, there exists an arc (x_i, x_j) with $x_i \in X_1$ and $x_j \in X_2$.

2.4. Prove that in an acyclic graph, the rank of a node is equal to the maximum number of arcs in a path terminating on that node.

2.5. Prove that in a complete binary tree, the number of terminal nodes is one more than the number of non-terminal nodes.

2.6. Let G be a tree, equipped with a transverse ordering $\leqslant$. Prove that if x_i and x_j are nodes of G, the node x_i is an ascendant of x_j if and only if x_i precedes x_j in the Tarry order and x_i follow x_j in the reverse Tarry order associated with $\leqslant$.

2.7. Given a graph $G = (X, U)$ we say that a subset Y of X is a *feedback node set* of G if every cycle of G traverses at least one node in Y.

(i) Develop a tree-search method to find the feedback node sets of a graph, and apply it to the graph of Fig. 2.1.
(ii) In practical situations we sometimes require only one feedback node set, of minimum cardinality. (For instance, in the diagnosis of faults in logic circuits, a feedback node set of minimum cardinality defines an appropriate set of 'test points' at which to monitor a circuit's behaviour.) Modify your search method, to determine such a set as efficiently as possible.

Additional notes and bibliography

The works of Ore (1962) and Berge (1976) are important reference texts on graph theory, giving a rigorous treatment of the subject; for an interesting historical account of its development, see Biggs, Lloyd, and Wilson (1976). Guidance on the choice of data structures and the development of algorithms for manipulating graphs on computers is given by Knuth (1968) and Aho, Hopcroft, and Ullman (1974).

The use of trees in sorting and searching is discussed in detail by Knuth (1973).

For general discussions of backtrack programming techniques, with examples, see Walker (1960), Golomb and Baumert (1965), Wells (1971), and Fillmore and Williamson (1974). Floyd (1967) describes techniques for developing backtrack algorithms.

Backtrack programming is essentially recursive, and backtrack algorithms can be defined very elegantly in recursive form; see Aho, Hopcroft, and Ullman (1974), and also Tarjan (1972). Bitner and Reingold (1975) discuss the implementation of backtrack algorithms using macros.

With reference to the examples used in Section 2.7, a backtrack algorithm for finding elementary paths has been published by Kroft (1967); it is simpler than the algorithm of Section 2.7, but it is usually much less efficient. The problem of finding all the elementary cycles of a graph has been studied extensively, because knowledge of the cycles is useful in optimizing computer programs; backtrack algorithms for finding the cycles (rather similar in principle to the algorithm of Section 2.7) have been developed by Tiernan (1970), Tarjan (1973), Johnson (1975), Read and Tarjan (1975), Tsukiyama, Shirakawa, and Ozaki (1975), and Szwarcfiter and Lauer (1976). The generation of Hamiltonian cycles by backtrack programming is discussed by Roberts and Flores (1966) and Rubin (1974). Kaufmann and Pichat (1977) have made a general study of path algorithms using stacks.

Tree-search methods have been extended to the *branch-and-bound* methods of solving discrete optimization problems (Lawler and Wood 1966; Mitten 1970; Garfinkel and Nemhauser 1972). Branch-and-bound methods of solving the travelling-salesman problem in particular are described by Little, Murty, Sweeney, and Karel (1963) and Bellmore and Nemhauser (1968); algorithms for finding a shortest Hamiltonian circuit on a simple graph are given by Held and Karp (1970, 1971). See also Bellmore and Hong (1974).

For a lucid introduction to the subject of computational complexity see Lawler (1976*a*), and for a more detailed treatment see Aho, Hopcroft, and Ullman (1974). The complexity of graph-theoretic problems has been studied extensively by Karp (1972, 1975*a,b*). See also Corneil (1974) and Knuth (1975).

Feedback node sets (as defined in Exercise 2.7) are also important in the formal verification of computer programs; backtrack programming methods for obtaining these sets are described by Guardabassi (1971) and Smith and Walford (1975).

3 Path problems

3.1. Introduction

PROBLEMS involving the determination of paths take many different forms. For instance, we have already encountered the problem of finding the 'critical' or longest paths in an activity graph. Later we shall present transportation problems in which we require 'least-cost' or shortest paths through a network, from the points where a commodity is produced to the points where it is consumed. Again, in transmitting messages through a communication network we may have to find a path of maximum reliability between two points, given the reliabilities (that is, the probabilities of successful operation) of the individual links.

These are all examples of 'extremal' path problems, i.e. problems in which each arc of a graph has a real number associated with it (representing for instance an activity duration, transit cost, or reliability), and in which we seek a path for which some function of the arc parameters is either maximized or minimized. Sometimes we also encounter path enumeration problems—such as the problem of finding all the elementary paths from one node to another, which arises for instance in testing computer logic circuits.

In this chapter we shall first present an algebraic structure—a 'path algebra'—which can be used to formulate and solve a wide variety of path problems. We shall then consider graphs whose arcs are 'labelled' with elements of a path algebra, which may be for instance real numbers, or words on some alphabet. It will then be shown that many path problems—including all those mentioned above—can be posed as a problem of solving a set of simultaneous equations in a path algebra. We shall then derive some direct and iterative methods of solving such equations, thereby obtaining different path-finding algorithms.

3.2. An algebra for path problems

3.2.1. Definition of a path algebra

We define a *path algebra* as a set P equipped with two binary operations $\vee$ and $\cdot$ which have the following properties.

(i)   *The $\vee$ operation is idempotent, commutative, and associative:*

$$x \vee x = x \qquad\qquad \textit{for all } x \in P, \tag{3.1}$$

$$x \vee y = y \vee x \qquad\qquad \textit{for all } x, y \in P, \tag{3.2}$$

$$(x \vee y) \vee z = x \vee (y \vee z) \quad \textit{for all } x, y, z \in P. \tag{3.3}$$

(ii) *The $\cdot$ operation is associative, and distributive over $\vee$:*

$$(x \cdot y) \cdot z = x \cdot (y \cdot z) \qquad\qquad \textit{for all } x, y, z \in P, \tag{3.4}$$

$$\begin{aligned} x \cdot (y \vee z) &= (x \cdot y) \vee (x \cdot z) \\ (y \vee z) \cdot x &= (y \cdot x) \vee (z \cdot x) \end{aligned} \qquad \textit{for all } x, y, z \in P. \tag{3.5}$$

(iii) *The set P contains a zero element ϕ such that*

$$\phi \vee x = x \qquad\qquad \textit{for all } x \in P, \tag{3.6}$$

$$\phi \cdot x = \phi = x \cdot \phi \quad \textit{for all } x \in P, \tag{3.7}$$

and a unit element e such that

$$e \cdot x = x = x \cdot e \qquad \textit{for all } x \in P. \tag{3.8}$$

The operation $\vee$ is called the *join operation* of P, and an element $x \vee y$ is called the *join* of x and y. The operation $\cdot$ is called *multiplication*, and an element $x \cdot y$ is described as the *product* of x and y (in that order). For simplicity, we may denote a product $x \cdot y$ by xy.

3.2.2. Some examples of the path algebra

Eight concrete examples of the path algebra are given in Table 3.1. For each example this table defines the set P, its join and multiplicative operations, and the zero and unit elements of P. The practical applications, which are also indicated in the table, will be discussed in more detail later in this chapter.

The algebra P_1 will be recognised as the two-element Boolean algebra. (In fact, any distributive lattice which has least and greatest elements can be regarded as a path algebra, the meet operation of the lattice playing the role of multiplication.)

The examples P_2–P_5 arise in connection with 'extremal' path problems of the kind described in the introduction. In P_2 for instance, the set P is the set $\mathbb{R}$ of real numbers, augmented by an element '∞'; the join operation is defined by

$$x \vee y = \min \{x, y\} \quad \textit{for all } x, y \in \mathbb{R},$$

$$x \vee \infty = x \qquad\qquad \textit{for all } x \in P,$$

TABLE 3.1

	P	$x \vee y$	$x \cdot y$	ϕ	e	Application
P_1:	$\{0, 1\}$	$\max\{x, y\}$	$\min\{x, y\}$	0	1	Determination of accessible sets
P_2:	$\mathbb{R} \cup \{\infty\}$	$\min\{x, y\}$	$x + y$	∞	0	Determination of shortest paths
P_3:	$\mathbb{R} \cup \{-\infty\}$	$\max\{x, y\}$	$x + y$	$-\infty$	0	Critical (longest) paths
P_4:	$\{x \in \mathbb{R} \mid 0 \leq x \leq 1\}$	$\max\{x, y\}$	$x \times y$	0	1	Most reliable paths
P_5:	$\{x \in \mathbb{R} \mid x \geq 0\} \cup \{\infty\}$	$\max\{x, y\}$	$\min\{x, y\}$	0	∞	Paths of greatest capacity
P_6:	$\mathscr{P}(\Sigma^*)$	$x \cup y$	$\{\chi \circ \psi \mid \chi \in x, \psi \in y\}$	ϕ	Λ	Listing all paths
P_7:	$\mathscr{P}(S)$	$x \cup y$	$\{\chi \circ \psi \in S \mid \chi \in x, \psi \in y\}$	ϕ	Λ	Listing simple paths
P_8:	B	$b\,(x \cup y)$	$\{\chi \circ \psi \mid \chi \in x, \psi \in y\}$	ϕ	Λ	Listing elementary paths

and the multiplicative operation is defined by

$$x \cdot y = x + y \qquad \text{for all } x, y \in \mathbb{R},$$

$$x \cdot \infty = \infty = \infty \cdot x \qquad \text{for all } x \in P.$$

It is easily verified that P obeys all the laws (3.1)–(3.8), the zero of P being ∞, the unit element being the number 0.

The algebras P_6–P_8 are derived from the linguistic concepts introduced in Example 1.23, as follows:

(i) Let Σ be any alphabet, let Σ^* be the set of all words over Σ, and let $\mathscr{P}(\Sigma^*)$ be the power set of Σ^*. (We recall that the elements of $\mathscr{P}(\Sigma^*)$, which are the subsets of Σ^*, are called *languages* over Σ.) For any two languages $X, Y \in \mathscr{P}(\Sigma^*)$, we define the *join* $X \vee Y$ by

$$X \vee Y = X \cup Y$$

and we define the *product $X \cdot Y$* by

$$X \cdot Y = \{ \chi \circ \psi \,|\, \chi \in X \text{ and } \psi \in Y \}$$

where $\chi \circ \psi$ denotes the concatenation of the words χ and ψ. (As an illustration, if $X = \{\lambda, a, ba\}$ and $Y = \{aa, b\}$ then $X \cdot Y = \{aa, b, aaa, ab, baaa, bab\}$.)

It is easily verified that the set $\mathscr{P}(\Sigma^*)$ with these operations is a path algebra, whose zero element is the *null language* (or null set) ϕ, and whose unit element is the language $\Lambda = \{\lambda\}$, where λ is the empty word. (This is the algebra P_6 of Table 3.1.)

(ii) Again, let Σ be any alphabet. Then we say that a word w on Σ is *simple* if no letter of Σ appears in w more than once. Let us denote the set of all simple words over Σ by S, and let $\mathscr{P}(S)$ be the power set of S, i.e. the set of all languages which comprise only simple words. For any two languages $X, Y \in \mathscr{P}(S)$ we define the *join* of X and Y as their set union:

$$X \vee Y = X \cup Y$$

and we define their *product $X \cdot Y$* as

$$X \cdot Y = \{ \chi \circ \psi \in S \,|\, \chi \in X \text{ and } \psi \in Y \}$$

where $\chi \circ \psi$ is the concatenation of the words χ and ψ. The set $\mathscr{P}(S)$ with these operations is a path algebra, whose zero and unit elements are the languages ϕ and Λ respectively. (This is the algebra P_7 of Table 3.1.)

(iii) We define an *abbreviation* of a word w as any word which can be obtained by removing at least one (and possibly all) of the letters of w (note that every word with at least one letter has the abbreviation λ). For any language X, we say that a word $w \in X$ is *basic to X* if X does not contain any abbreviation of w, and we describe the set $b(X)$ of all basic words of X as the *basis* of X. If $b(X) = X$ then X is a *basic language*; in particular, the languages ϕ and Λ are both basic.

Now let Σ be any alphabet, and let B be the set of all basic languages on Σ. For any two languages X, $Y \in B$, we define the *join* of X and Y by

$$X \vee Y = b(X \cup Y)$$

and we define the *product* $X \cdot Y$ by

$$X \cdot Y = \{\chi \circ \psi \,|\, \chi \in X \text{ and } \psi \in Y\}.$$

The set B with these operations is a path algebra, with zero and unit elements ϕ and Λ respectively. (This is the algebra P_8 of Table 3.1.)

Some further examples of path algebras will be given later in this chapter, and in Chapter 4.

3.2.3. Elementary properties of path algebras

Since the join operation of P is idempotent, commutative, and associative, we can define an ordering $\leqslant$ of P by the following rule (see Section 1.5.2):

$$x \leqslant y \quad \text{if and only if} \quad x \vee y = y. \tag{3.9}$$

It is evident from (3.6) that with respect to this ordering, ϕ is the least element of P:

$$\phi \leqslant x \quad \text{for all } x \in P. \tag{3.10}$$

Also, since the join operation is idempotent,

$$x \vee y \geqslant x \quad \text{and} \quad x \vee y \geqslant y \quad \text{for all } x, y \in P. \tag{3.11}$$

It is important also to note that the join operation is isotone for $\leqslant$:

$$\text{if} \quad x \leqslant y \quad \text{then} \quad x \vee z \leqslant y \vee z \quad \text{for all } z \in P. \tag{3.12}$$

(For the proof, see Section 1.5.4.) The multiplicative operation is also isotone for $\leqslant$:

$$\text{if} \quad x \leqslant y \quad \text{then} \quad x \cdot z \leqslant y \cdot z \quad \text{and} \quad z \cdot x \leqslant z \cdot y \quad \text{for all } z \in P. \tag{3.13}$$

For if $x \leqslant y$ then $x \vee y = y$ and therefore (by the distributive law (3.5)) $y \cdot z = (x \vee y) \cdot z = (x \cdot z) \vee (y \cdot z)$, which implies that $x \cdot z \leqslant y \cdot z$; we obtain $z \cdot x \leqslant z \cdot y$ by a similar argument.

In a path algebra, an element x is said to be *sub-unitary* if $x \leqslant e$, whereas x is said to be *super-unitary* if $x \geqslant e$.

Example 3.1. For the path algebras P_1, P_3, P_4 and P_5 the ordering $\leqslant$ is the familiar ordering $\leq$ of the real numbers, whereas for P_2 the ordering $\leqslant$ is the familiar ordering $\geq$. For P_6 and P_7 $\leqslant$ becomes the set inclusion relation $\subseteq$; while in P_8, $x \leqslant y$ means that each word in the language x occurs in the language y, or has an abbreviation in y.

Powers. The *powers* of an element $x \in P$ are defined by

$$x^0 = e, \qquad x^k = x^{k-1}x \qquad (k = 1, 2, \ldots). \tag{3.14}$$

An element x for which $x^2 = x$ is said to be *idempotent*. An element x such that $x^q = \phi$ for some positive integer q is said to be *nilpotent*.

Closure. An element x is said to be *stable* if for some non-negative integer q,

$$\bigvee_{k=0}^{q} x^k = \bigvee_{k=0}^{q+1} x^k \tag{3.15}$$

where $\bigvee_{k=0}^{q} x^k$ denotes the join $e \vee x \vee x^2 \vee \cdots \vee x^q$; the least value of q for which (3.15) holds is called the *stability index* of x.

If (3.15) holds then, by multiplying both sides by x, we have

$$\bigvee_{k=1}^{q+1} x^k = \bigvee_{k=1}^{q+2} x^k$$

and by joining x^0 to each side we obtain

$$\bigvee_{k=0}^{q+1} x^k = \bigvee_{k=0}^{q+2} x^k. \tag{3.16}$$

From (3.15) and (3.16),

$$\bigvee_{k=0}^{q} x^k = \bigvee_{k=0}^{q+2} x^k.$$

By repetition of this argument it follows that if x is stable of index q then

$$\bigvee_{k=0}^{q} x^k = \bigvee_{k=0}^{r} x^k \quad \text{for all } r \geq q. \tag{3.17}$$

For a stable element x, the join (3.17) is called the *strong closure* (or simply the *closure*) of x, and it is denoted by x^*.

As an important special case, it will be observed that if an element x is sub-unitary then $x^k \leqslant e$, $(k = 1, 2, \ldots)$, and therefore x is stable of index 0, with

$$x^* = e. \tag{3.18}$$

In particular, the zero element ϕ of any path algebra is stable, with closure

$$\phi^* = e. \tag{3.19}$$

Example 3.2. In the path algebras P_1, P_4, P_5, and P_8 of Table 3.1, the unit element is the greatest element. Hence in those algebras all elements are stable, and their closures are given by (3.18).

Example 3.3. In a path algebra whose multiplicative operation has the cancellation property, an element x is stable if and only if it is sub-unitary. Indeed, it has already been shown that if $x \leqslant e$ then x is stable. To prove the converse, let us assume that x is stable; then, since $x^* \vee xx^* = x^*$,

$$(e \vee x)x^* = ex^*.$$

It follows (by cancellation) that $e \vee x = e$, which implies that x is sub-unitary.

As concrete examples, in both P_2 and P_3 the multiplicative operation has the cancellation property. In P_2 the non-negative numbers are stable (of index 0) and the negative numbers are unstable, whereas in P_3 the non-positive numbers are stable and the positive numbers are unstable.

Example 3.4. Let x be any element of the algebra P_7. Then x^k is the set of all simple words which can be obtained by concatenating k words of x. For any non-negative integer r,

$$\bigvee_{k=0}^{r} x^k = \bigcup_{k=0}^{r} x^k,$$

which is the set comprising the empty word λ, the words of x, and all the simple words which can be obtained by concatenating up to r words in x. Now let q be the number of letters of the alphabet Σ. Then, since no simple word has more than q letters,

$$\bigvee_{k=0}^{q} x^k = \bigvee_{k=0}^{r} x^k \quad \text{for all } r \geq q.$$

It follows that all elements of P_7 are stable.

From the definition of closure, it follows immediately that for any stable element x,

$$x^* = e \vee xx^*, \tag{3.20}$$

$$x^* = (e \vee x)^*, \tag{3.21}$$

$$(x^*)^* = x^*, \tag{3.22}$$

$$x^*x^* = x^*. \tag{3.23}$$

Also, provided that the 'starred' terms are stable,

$$(xy)^* = e \vee x(yx)^*y, \tag{3.24}$$

$$(xy)^*x = x(yx)^*, \tag{3.25}$$

$$(x \vee y)^* = (x^*y^*)^*, \tag{3.26}$$

$$(x \vee y)^* = (x^*y)^*x^*, \tag{3.27}$$

$$(x \vee y)^* = x^*(yx^*)^*. \tag{3.28}$$

Indeed,

$$(xy)^* = e \vee xy \vee x(yx)y \vee x(yxyx)y \vee \cdots = e \vee x(yx)^*y$$

proving (3.24), and similarly

$$(xy)^*x = x \vee (xy)x \vee (xyxy)x \vee \cdots$$
$$= x \vee x(yx) \vee x(yxyx) \vee \cdots$$
$$= x(yx)^*,$$

which proves (3.25). For the identity (3.26) it can be shown by expansion that $(x \vee y)^*$ is the join of all products of x's and y's, and that $(x^*y^*)^*$ is also the join of these terms. A similar argument is used to prove (3.27) and (3.28).

Finally, we note that by (3.12) and (3.13),

$$x \leqslant y \quad \text{implies that} \quad x^* \leqslant y^* \tag{3.29}$$

provided of course that both x and y are stable.

Weak closure. From (3.17) it follows that for any element x which is stable of index q,

$$\bigvee_{k=1}^{q+1} x^k = \bigvee_{k=1}^{r} x^k \quad \text{for all } r \geqslant q+1. \tag{3.30}$$

The join (3.30) is called the *weak closure* $\hat{x}$ of x. It is evident that

$$\hat{x} = xx^* = x^*x, \tag{3.31}$$

$$x^* = e \vee \hat{x}. \tag{3.32}$$

3.2.4. The solution of equations

Let us now consider the problem of finding an element y which satisfies an equation

$$y = ay \vee b \tag{3.33}$$

where a and b are specified, the element a being stable of index q.

First, it will be observed that the equation has a solution

$$y = a^*b, \tag{3.34}$$

for it follows from (3.20) that

$$a(a^*b) \vee b = (aa^* \vee e)b = a^*b.$$

Now let y_0 be an arbitrary solution of (3.33), i.e.

$$y_0 = ay_0 \vee b.$$

By substituting, we obtain

$$y_0 = a(ay_0 \vee b) \vee b = a^2y_0 \vee (e \vee a)b,$$

and by repeated substitutions

$$y_0 = a^k y_0 \vee (e \vee a \vee a^2 \vee \cdots \vee a^{k-1})b, \qquad (k = 1, 2, \ldots).$$

It follows that

$$y_0 = a^k y_0 \vee a^*b \quad \text{for all } k > q, \tag{3.35}$$

and consequently (by (3.11)),

$$y_0 \geqslant a^*b. \tag{3.36}$$

From this inequality it follows that the solution (3.34) is the *least* solution of the equation (3.33). It is also evident from (3.35) that in the particular case where a is nilpotent, the solution (3.34) is *unique*.

By a similar argument it can be shown that if an element a is stable then the equation

$$y = ya \vee b \tag{3.33'}$$

has the least solution

$$y = ba^*, \tag{3.34'}$$

and again, if a is nilpotent this solution is unique.

Example 3.5. If the element a in equation (3.33) is sub-unitary (as is always the case for instance in a Boolean algebra), then by (3.18) $a^* = e$, and the least solution (3.34) becomes $y = b$. This solution is not necessarily unique: for instance, it is easily verified that in any Boolean algebra, the equation $y = ay \vee b$ has for solutions all those elements y which satisfy the condition $b \leqslant y \leqslant a \vee b$.

Example 3.6. Let us consider the equation (3.33) in the algebra P_7, in the particular case in which $a = \{\alpha\}$ and $b = \{\beta\}$, where α and β are two non-empty words which do not have any letters in common. Here a is nilpotent, with $a^2 = \phi$, hence the equation (3.33) has a unique solution

$$y = a^*b = (\Lambda \vee a)b = \{\lambda, \alpha\} \cdot \{\beta\} = \{\beta, \alpha\beta\}.$$

3.2.5. Matrices

Let P be a path algebra, and let $M_n(P)$ be the set of all $n \times n$ matrices whose entries belong to P. We define two binary operations on $M_n(P)$ as follows: given any matrices $X = [x_{ij}]$ and $Y = [y_{ij}]$ in $M_n(P)$, their *join* is the $n \times n$ matrix

$$X \vee Y = [x_{ij} \vee y_{ij}] \tag{3.37}$$

and their *product* XY is the $n \times n$ matrix

$$XY = \left[\bigvee_{k=1}^{n} x_{ik} \cdot y_{kj} \right]. \tag{3.38}$$

Thus we form the join $X \vee Y$ and the product XY in the same way as we construct the sum and product in ordinary matrix algebra, except that here we use the join operation of P in place of addition on $\mathbb{R}$, and we use multiplication on P in place of multiplication on $\mathbb{R}$.

It is easy to establish that *for any path algebra P, the set of matrices $M_n(P)$, equipped with the join and multiplicative operations defined in (3.37) and (3.38), is itself a path algebra.*

Indeed, it follows immediately from the definition (3.37) that the join operation of $M_n(P)$ is idempotent, commutative and associative. Also, since in P multiplication is associative and distributive over the join operation, matrix multiplication is also associative, and distributive over the matrix join operation. (These properties can be demonstrated by the same arguments as are used in ordinary matrix algebra to prove that matrix multiplication is associative, and distributive over matrix addition—see for instance Birkhoff and MacLane (1965).) Finally, the $n \times n$ matrix Φ whose entries are all ϕ (the zero element of P) satisfies the conditions (cf. (3.6) and (3.7))

$$\Phi \vee X = X \quad \text{and} \quad \Phi X = \Phi = X\Phi \quad \text{for all } X \in M_n(P)$$

and the $n \times n$ matrix

$$E = \begin{bmatrix} e & \phi & \phi & \cdots & \phi \\ \phi & e & \phi & \cdots & \phi \\ \phi & \phi & e & \cdots & \phi \\ \cdots\cdots\cdots\cdots\cdots\cdots \\ \phi & \phi & \phi & \cdots & e \end{bmatrix}$$

is a unit element for multiplication (cf. (3.8)):

$$EX = X = XE \quad \text{for all } X \in M_n(P).$$

The matrices Φ and E are called respectively the *zero matrix* and the *unit matrix* of $M_n(P)$.

Since the matrix algebra $M_n(P)$ is a path algebra, we can extend all the definitions and results of Sections 3.2.3 and 3.2.4 to $M_n(P)$, without modification.

Thus, we may define an ordering of matrices by the rule (cf. (3.9))

$$X \leqslant Y \quad \text{if and only if} \quad X \vee Y = Y.$$

From (3.9) and (3.37) it follows that this ordering of matrices can be expressed in terms of the ordering of their entries:

$$X \leqslant Y \quad \text{if and only if} \quad x_{ij} \leqslant y_{ij} \quad \text{for all } i, j.$$

The significance of closure for matrices, and the solution of matrix equations, will be discussed later in this chapter.

Example 3.7. As a concrete example, let us take the path algebra P_2 of Table 3.1. The operations of the matrix algebra $M_n(P_2)$ are defined by

$$X \vee Y = [\min\{x_{ij}, y_{ij}\}] \quad \text{and} \quad XY = \left[\min_{1 \leq k \leq n} \{x_{ik} + y_{kj}\} \right],$$

$$\text{for all } X, Y \in M_n(P_2).$$

The zero matrix Φ of $M_n(P_2)$ is the $n \times n$ matrix all of whose entries are ∞, while the unit matrix $E = [e_{ij}]$ has $e_{ij} = 0$ if $i = j$ and $e_{ij} = \infty$ if $i \neq j$. As a numerical illustration,

$$\text{for} \quad X = \begin{bmatrix} 0 & \infty \\ 1 & 2 \end{bmatrix} \quad \text{and} \quad Y = \begin{bmatrix} 3 & 4 \\ \infty & 5 \end{bmatrix},$$

$$X \vee Y = \begin{bmatrix} 0 & 4 \\ 1 & 2 \end{bmatrix} \quad \text{and} \quad XY = \begin{bmatrix} 3 & 4 \\ 4 & 5 \end{bmatrix}.$$

3.3. Labelled graphs

3.3.1. *Definition of a labelled graph*

A graph $G = (X, U)$ is said to be *labelled* with a path algebra P when each arc of G is assigned some element of P, other than its zero element ϕ. We describe the element of P which is assigned to an arc (x_i, x_j) as its *label*, and denote it by $l(x_i, x_j)$.

3.3.2. *Path labels*

Let G be a graph labelled with a path algebra P, and let $\mu = (x_{i_0}, x_{i_1}), (x_{i_1}, x_{i_2}), \ldots, (x_{i_{r-1}}, x_{i_r})$ be any path on G. Then the *label* $l(\mu)$ *of the path* μ is defined as the product of the arc labels of μ, taken in order:

$$l(\mu) = l(x_{i_0}, x_{i_1}) \cdot l(x_{i_1}, x_{i_2}) \cdot \cdots \cdot l(x_{i_{r-1}}, x_{i_r}).$$

In the present context, it will be convenient to suppose that each node x_i of a graph G is connected to itself by a *null path* θ_i. By definition, each null path θ_i is of order zero (for it contains no arcs), and $l(\theta_i) = e$.

Example 3.8. In graphs labelled with the algebras P_2–P_5 of Table 3.1, each arc label is a real number. If a graph G is labelled with P_2 or P_3—as in Fig. 3.1 for example—then the label $l(\mu)$ of a path μ on G is the

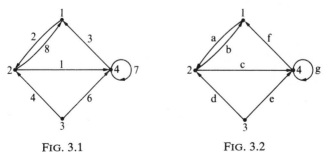

FIG. 3.1   FIG. 3.2

arithmetical sum of the labels on the arcs of μ. (Thus if the labels on arcs represent their lengths, as on a road map for instance, the label $l(\mu)$ is the total length of μ.) If G is labelled with P_4, $l(\mu)$ is the arithmetical product of the labels of the arcs of μ. (In this way, if G represents a communication network, and each arc label represents its reliability, the label $l(\mu)$ is the reliability of the path μ.) If G is labelled with P_5 then $l(\mu)$ is the smallest of the labels on the arcs of μ. (Hence if each arc label represents its capacity, i.e. the rate at which some substance can flow along the arc, then $l(\mu)$ is the capacity of the path μ.)

Example 3.9. Fig. 3.2 shows a graph whose arcs have distinct names a, b, c, ... Now this graph can be considered being labelled with the linguistic algebra P_6, where $\Sigma = \{a, b, c, \ldots\}$. (To be precise, the labels assigned to arcs are the one-word languages {a}, {b}, {c} . . . which comprise their names.) Then for any path μ, the label $l(\mu)$ is a language comprising a single word, this word being the concatenation of the names of the arcs of μ. For instance, the path $\mu = (x_3, x_2), (x_2, x_4), (x_4, x_1)$ on Fig. 3.2 has the label

$$l(\mu) = \{d\} \cdot \{c\} \cdot \{f\} = \{dcf\}.$$

Since the word in the label $l(\mu)$ of a path μ completely defines μ, we may describe this word as the *name* of μ.

3.3.3. Absorptive graphs

A graph G labelled with a path algebra P is said to be *absorptive* if for every elementary cycle γ on G,

$$l(\gamma) \leqslant e. \tag{3.39}$$

Let G be an absorptive graph, and let us suppose that G contains a non-elementary path μ, from x_i to x_j say. (Here the nodes x_i and x_j need not be distinct.) Since μ is non-elementary, it traverses at least one elementary cycle. We may therefore regard μ as the concatenation of three paths β_1, γ_1, δ_1, where γ_1 is an elementary cycle, and where the paths β_1 and δ_1 together form a single path μ_1 from x_i to x_j, which is of lower order than μ. (It is possible for either β_1 or δ_1 to be a null path.) Now since the multiplicative operation of P is associative, we can write the labels of μ and μ_1 as

$$l(\mu) = l(\beta_1) \cdot l(\gamma_1) \cdot l(\delta_1), \tag{3.40}$$

and

$$l(\mu_1) = l(\beta_1) \cdot l(\delta_1), \tag{3.41}$$

and since G is absorptive it follows from (3.39), (3.40), and (3.41) that

$$l(\mu) \leqslant l(\mu_1). \tag{3.42}$$

If μ_1 is non-elementary, we may also consider this path as the concatenation of three paths β_2, γ_2, δ_2, where γ_2 is an elementary cycle and the segments β_2 and δ_2 together form a path μ_2 from x_i to x_j, of lower order than μ_1. Repeating the previous argument, we find that $l(\mu_1) \leqslant l(\mu_2)$, and since the relation $\leqslant$ is transitive it follows by (3.42) that

$$l(\mu) \leqslant l(\mu_2). \tag{3.43}$$

By repetition of this process, we must eventually obtain an elementary path μ_k from x_i to x_j, such that $l(\mu) \leqslant l(\mu_k)$. To summarize:

In any absorptive graph, if there exists a non-elementary path μ from a node x_i to a node x_j, then there also exists an elementary path $\tilde{\mu}$ from x_i to x_j, of order less than μ, and such that $l(\mu) \leqslant l(\tilde{\mu})$.

Example 3.10. Let G be a graph labelled with P_2, and let us interpret the (numerical) labels of the arcs of G as their lengths. Then G is absorptive if and only if the lengths of its elementary cycles are all non-negative. For the case where G is absorptive our theorem above states that if there exists a non-elementary path μ from x_i to x_j, then there also exists an elementary path from x_i to x_j, whose length is less than or equal to the length of μ.

3.4. Graphs and matrices

3.4.1. Adjacency matrices

An n-node graph $G = (X, U)$ labelled with a path algebra P can be described by its *adjacency matrix*, which is the $n \times n$ matrix $A = [a_{ij}]$ of $M_n(P)$ with entries

$$a_{ij} = \begin{cases} l(x_i, x_j) & \text{if } (x_i, x_j) \in U, \\ \phi & \text{if } (x_i, x_j) \notin U, \end{cases}$$

where ϕ is the zero element of P. Conversely, any $n \times n$ matrix A whose entries all belong to a path algebra P can be visualized as an n-node labelled graph $G = (X, U)$ where $U = \{(x_i, x_j) \in X \times X \mid a_{ij} \neq \phi\}$, each arc (x_i, x_j) of G having a label $l(x_i, x_j) = a_{ij}$.

Example 3.11. In the case where a graph $G = (X, U)$ is labelled with the two-element Boolean algebra, every arc has the label '1' (since the zero element of a path algebra cannot be used as an arc label—see Section 3.3.1), and the adjacency matrix A of G has

$$a_{ij} = \begin{cases} 1 & \text{if } (x_i, x_j) \in U, \\ 0 & \text{if } (x_i, x_j) \notin U. \end{cases}$$

This matrix is usually called the *Boolean adjacency matrix* of G. As an illustration, the Boolean adjacency matrix of the graph cf Fig. 3.3(a) is given in Fig. 3.3(b).

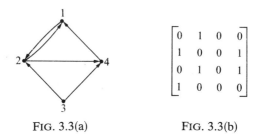

$$\begin{bmatrix} 0 & 1 & 0 & 0 \\ 1 & 0 & 0 & 1 \\ 0 & 1 & 0 & 1 \\ 1 & 0 & 0 & 0 \end{bmatrix}$$

FIG. 3.3(a)   FIG. 3.3(b)

In Section 2.2 we defined the *complement* $\bar{G}$ and the *converse* G' of a graph G. It is easily verified that if G has a Boolean adjacency matrix A then the adjacency matrix of $\bar{G}$ is $\bar{A} = [\bar{a}_{ij}]$, where $\bar{a}_{ij}$ denotes the Boolean complement of a_{ij}; the adjacency matrix of G' is the *transpose* A' of A.

As one might expect, many basic properties of graphs can be interpreted very simply in terms of their Boolean adjacency matrices. For instance, given a graph G with adjacency matrix A: (i) G is *reflexive* if and only if $E \leqslant A$; (ii) G is *symmetric* if and only if $A = A'$; and (iii) G is *transitive* if and only if $A^2 \leqslant A$.

Example 3.12. The graph of Fig. 3.1, labelled with P_2, has the adjacency matrix

$$A = \begin{bmatrix} \infty & 2 & \infty & \infty \\ 8 & \infty & \infty & 1 \\ \infty & 4 & \infty & 6 \\ 3 & \infty & \infty & 7 \end{bmatrix}.$$

Adjacency matrices of acyclic graphs. In Section 2.9 it was shown that if the nodes of an acyclic graph $G = (X, U)$ are numbered in order of increasing rank, then for every arc (x_i, x_j) in G we have $i < j$.

In this case, the adjacency matrix A of G has $a_{ij} = \phi$ whenever $i \geqslant j$; in other words, all elements on and below the principal diagonal are null. We describe such a matrix as being *strictly upper triangular*. In the same way, if the nodes of a graph are numbered in order of *decreasing* rank, its adjacency matrix is *strictly lower triangular*.

Conversely, it follows directly from the characterization of acyclic graphs in Section 2.5 that if a matrix $A \in M_n(P)$ is strictly upper or lower triangular, then its labelled graph is acyclic.

As an illustration, Fig. 3.4 shows a graph whose nodes are numbered in order of increasing rank, together with its Boolean adjacency matrix.

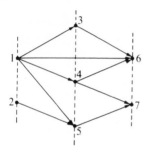

$$\begin{bmatrix} 0 & 0 & 1 & 1 & 1 & 1 & 0 \\ 0 & 0 & 0 & 0 & 1 & 0 & 0 \\ 0 & 0 & 0 & 0 & 0 & 1 & 0 \\ 0 & 0 & 0 & 0 & 0 & 1 & 1 \\ 0 & 0 & 0 & 0 & 0 & 0 & 1 \\ 0 & 0 & 0 & 0 & 0 & 0 & 0 \\ 0 & 0 & 0 & 0 & 0 & 0 & 0 \end{bmatrix}$$

FIG. 3.4(a)   FIG. 3.4(b)

3.4.2. Powers of matrices

Let P be any path algebra, and let A be any matrix belonging to $M_n(P)$. Then, by the definition (3.14), the powers of A are

$$A^0 = E, \qquad A^k = A^{k-1}A \qquad (k = 1, 2, \ldots),$$

where E is the unit matrix of $M_n(P)$.

The elements of these powers can be defined in terms of the labels of paths on the graph corresponding to A, as follows:

Let a_{ij}^k denote the typical element of A^k, and let S_{ij}^k be the set of all paths of order k from x_i to x_j, on the labelled graph G of A. Then

$$a_{ij}^k = \bigvee_{\mu \in S_{ij}^k} l(\mu), \qquad (k = 0, 1, 2, \ldots).\dagger \qquad (3.44)$$

$\dagger$ To simplify notation, we adopt the usual convention that $\bigvee_{x \in S} x$ denotes ϕ if S is empty.

This formula is obviously valid for $k = 0$, since $S_{ii}^0 = \{\theta_i\}$, ($i = 1, 2, \ldots, n$), where θ_i is the null path associated with x_i, while S_{ij}^0 is empty if $i \neq j$. For $k > 0$, the typical element of A^k can be written as

$$a_{ij}^k = \bigvee_{h_0, h_1, \ldots, h_{k-1}} a_{ih_1} \cdot a_{h_1 h_2} \cdots a_{h_{k-1}j} \qquad (3.45)$$

where $\bigvee_{h_0, h_1, \ldots, h_{k-1}}$ denotes the join extended over the n^{k-1} possible values of the indices. Since the label of every path in S_{ij}^k is a product in (3.45), and every non-null product in (3.45) is the label of a path in S_{ij}^k, the formula (3.44) is also valid for all $k > 0$.

Example 3.13. On a graph G labelled with the two-element Boolean algebra, every arc has the label 1, and every path μ has the label $l(\mu) = 1$. The powers A^k of the Boolean adjacency matrix of G have elements $a_{ij}^k = 1$ if G contains any paths of order k from x_i to x_j and $a_{ij}^k = 0$ otherwise.

For the graph of Fig. 3.3(a), whose adjacency matrix A is given in Fig. 3.3(b),

$$A^2 = \begin{bmatrix} 1 & 0 & 0 & 1 \\ 1 & 1 & 0 & 0 \\ 1 & 0 & 0 & 1 \\ 0 & 1 & 0 & 0 \end{bmatrix}, \quad A^3 = \begin{bmatrix} 1 & 1 & 0 & 0 \\ 1 & 1 & 0 & 1 \\ 1 & 1 & 0 & 0 \\ 1 & 0 & 0 & 1 \end{bmatrix}, \quad A^4 = \begin{bmatrix} 1 & 1 & 0 & 1 \\ 1 & 1 & 0 & 1 \\ 1 & 1 & 0 & 1 \\ 1 & 1 & 0 & 0 \end{bmatrix}.$$

Example 3.14. Let G be a graph labelled with P_2, and let us interpret its arc labels as physical lengths (see Examples 3.8, 3.10). Then each element a_{ij}^k is the length of a shortest path of order k from x_i to x_j, provided that these nodes are joined by at least one path of order k; otherwise $a_{ij}^k = \infty$.

For the graph of Fig. 3.1, whose adjacency matrix A is given in Example 3.12,

$$A^2 = \begin{bmatrix} 10 & \infty & \infty & 3 \\ 4 & 10 & \infty & 8 \\ 9 & \infty & \infty & 5 \\ 10 & 5 & \infty & 14 \end{bmatrix}, \quad A^3 = \begin{bmatrix} 6 & 12 & \infty & 10 \\ 11 & 6 & \infty & 11 \\ 8 & 11 & \infty & 12 \\ 13 & 12 & \infty & 6 \end{bmatrix}.$$

Example 3.15. Let G be a graph whose arcs have distinct names a, b, c, $\ldots$, and let us consider these names as elements of the algebra P_6 (as in Example 3.9). Let A be the adjacency matrix of G. Then each element a_{ij}^k is the set of names of all the paths of order k from x_i to x_j. Thus for the graph of Fig. 3.2,

$$
A = \begin{bmatrix} \phi & \{a\} & \phi & \phi \\ \{b\} & \phi & \phi & \{c\} \\ \phi & \{d\} & \phi & \{e\} \\ \{f\} & \phi & \phi & \{g\} \end{bmatrix}, \quad
A^2 = \begin{bmatrix} \{ab\} & \phi & \phi & \{ac\} \\ \{cf\} & \{ba\} & \phi & \{cg\} \\ \{db, ef\} & \phi & \phi & \{dc, eg\} \\ \{gf\} & \{fa\} & \phi & \{gg\} \end{bmatrix},
$$

$$
A^3 = \begin{bmatrix} \{acf\} & \{aba\} & \phi & \{acg\} \\ \{bab, cgf\} & \{cfa\} & \phi & \{bac, cgg\} \\ \{dcf, egf\} & \{dba, efa\} & \phi & \{dcg, egg\} \\ \{fab, ggf\} & \{gfa\} & \phi & \{fac, ggg\} \end{bmatrix}.
$$

For any matrix $A \in M_n(P)$ and any positive integer h, let us denote by $A^{[h]} = a_{ij}^{[h]}$ the join

$$
A^{[h]} = \bigvee_{k=0}^{h} A^k. \tag{3.46}
$$

Then $A^{[h]}$ has entries

$$
a_{ij}^{[h]} = \bigvee_{k=0}^{h} a_{ij}^k
$$

hence

$$
a_{ij}^{[h]} = \begin{cases} \displaystyle\bigvee_{k=1}^{h} a_{ij}^k & \text{if } i \neq j, \\[2ex] e \vee \left(\displaystyle\bigvee_{k=1}^{h} a_{ij}^k \right) & \text{if } i = j. \end{cases} \tag{3.47}
$$

By (3.44)

$$
\bigvee_{k=1}^{h} a_{ij}^k = \bigvee_{k=1}^{h} \bigvee_{\mu \in S_{ij}^k} l(\mu) = \bigvee_{\mu \in T_{ij}^h} l(\mu) \quad \text{for all } i, j \tag{3.48}
$$

where T_{ij}^h denotes the set of all non-null paths from x_i to x_j, of order less than or equal to h, and therefore (3.47) can be written as

$$
a_{ij}^{[h]} = \begin{cases} \displaystyle\bigvee_{\mu \in T_{ij}^h} l(\mu) & \text{if } i \neq j, \\[2ex] l(\theta_i) \vee \left(\displaystyle\bigvee_{\mu \in T_{ij}^h} l(\mu) \right) & \text{if } i = j. \end{cases} \tag{3.49}
$$

In words, each element $a_{ij}^{[h]}$ of $A^{[h]}$ in the join of the labels of all the paths from x_i to x_j, of order less than or equal to h.

Example 3.16. Let A be the Boolean adjacency matrix of a graph G. Then $A^{[h]}$ has entries $a_{ij}^{[h]} = 1$ if G has any paths from x_i to x_j of order less than or equal to h, and $a_{ij}^{[h]} = 0$ otherwise.

For the A-matrix of Fig. 3.3, whose powers were given in Example 3.13,

$$A^{[1]} = \begin{bmatrix} 1 & 1 & 0 & 0 \\ 1 & 1 & 0 & 1 \\ 0 & 1 & 1 & 1 \\ 1 & 0 & 0 & 1 \end{bmatrix} \quad \text{and} \quad A^{[h]} = \begin{bmatrix} 1 & 1 & 0 & 1 \\ 1 & 1 & 0 & 1 \\ 1 & 1 & 1 & 1 \\ 1 & 1 & 0 & 1 \end{bmatrix} \quad \text{for all } h \geq 2.$$

3.4.3. Stable matrices

We recall that since $M_n(P)$ is a path algebra, all the definitions and results of Sections 3.2.3 and 3.2.4 can be applied to matrices. Accordingly, we say that a matrix $A \in M_n(P)$ is *stable* if for some non-negative integer q,

$$\bigvee_{k=0}^{q} A^k = \bigvee_{k=0}^{q+1} A^k. \tag{3.50(a)}$$

or (using the notation of the previous section)

$$A^{[q]} = A^{[q+1]}, \tag{3.50(b)}$$

and we describe the least value of q for which (3.50) holds as the *stability index* of A.

The definitions of closures given in Section 3.2.3 will also be applied to matrices; we shall denote the strong and weak closure of a stable matrix A by $A^* = [a_{ij}^*]$ and $\hat{A} = [\hat{a}_{ij}]$ respectively. (To denote the strong closure of an *element* a_{ij} of A, we shall use the symbolism $(a_{ij})^*$.)

Since $A^* = E \vee \hat{A}$ (cf. (3.32)), the elements of A^* and $\hat{A}$ are related by

$$a_{ij}^* = \begin{cases} \hat{a}_{ij}, & \text{if } i \neq j, \\ e \vee \hat{a}_{ij} & \text{if } i = j. \end{cases} \tag{3.51}$$

Example 3.17. Let G be a graph with m arcs, which are assigned 'names', i.e. distinct symbols from an alphabet Σ, and let us consider these arc labels as elements of the path algebra P_7. Then for any simple path μ on G, $l(\mu) = \{\sigma\}$, where σ is the 'name' of μ (as defined in Example 3.9); whereas for any non-simple path μ, $l(\mu)$ is the empty set ϕ.

Now let us consider the adjacency matrix A of G. By (3.44) each element a_{ij}^k of A^k is the set of names of all the simple paths of order k from x_i to x_j.

Since no simple path has more than m arcs, $A^k = \Phi$ for all $k > m$, and therefore A is stable, with a stability index not greater than m. Each element a_{ij}^* of A^* is the set of names of all the simple paths from x_i to x_j (which in the case of a diagonal element includes the name λ of the null path θ_i). Each element $\hat{a}_{ij}$ of $\hat{A}$ is the set of names of all the non-null simple paths from x_i to x_j.

A particularly important class of matrices which have the stable property is presented in the next section.

3.4.4. *Absorptive matrices*

A matrix $A \in M_n(P)$ is said to be *absorptive* if its graph is absorptive.

Let A be an absorptive matrix. It was shown in Section 3.3.3 that if the graph G of A contains a non-elementary path μ from x_i to x_j, then G also contains an elementary path $\tilde{\mu}$ between these nodes, such that $l(\mu) \vee l(\tilde{\mu}) = l(\tilde{\mu})$. It follows that for any positive integer h,

$$\bigvee_{\mu \in T_{ij}^h} l(\mu) = \bigvee_{\mu \in \tilde{T}_{ij}^h} l(\mu) \quad \text{for all } i, j, \tag{3.52}$$

where $\tilde{T}_{ij}^h$ denotes the set of all non-null *elementary* paths from x_i to x_j, of order less than or equal to h.

Now since G has n nodes, each open elementary path on G is of order less than n, and each elementary cycle of G is of order less than or equal to n. Thus if we denote by $\tilde{T}_{ij}$ the set of all non-null elementary paths from x_i to x_j, we have by (3.52) that

$$\text{if } i \neq j \quad \text{then} \bigvee_{\mu \in T_{ij}^h} l(\mu) = \bigvee_{\mu \in \tilde{T}_{ij}} l(\mu) \quad \text{for all } h \geq n - 1,$$
$$\tag{3.53(a)}$$

$$\text{if } i = j \quad \text{then} \bigvee_{\mu \in T_{ij}^h} l(\mu) = \bigvee_{\mu \in \tilde{T}_{ij}} l(\mu) \quad \text{for all } h \geq n.$$
$$\tag{3.53(b)}$$

Also, it follows from (3.52) and (3.39) that for any positive integer h,

$$\bigvee_{\mu \in T_{ij}^h} l(\mu) = \bigvee_{\mu \in \tilde{T}_{ij}^h} l(\mu) \leqslant e \quad \text{if } i = j. \tag{3.54}$$

It follows from (3.49), (3.53(a)) and (3.54) that $A^{[h]}$ has entries

$$a_{ij}^{[h]} = \begin{cases} \bigvee_{\mu \in \tilde{T}_{ij}} l(\mu) & \text{if } i \neq j \\ e & \text{if } i = j \end{cases} \quad \text{for all } h \geq n - 1, \quad (3.55)$$

which implies that $A^{[h]} = A^{[n-1]}$, for all $h \geq n - 1$. Thus, *every absorptive matrix of $M_n(P)$ is stable, with a stability index not greater than $n - 1$.*

Since $A^* = A^{[n-1]}$, the entries of A^* are evidently given by (3.55). For the weak closure,

$$\hat{A} = AA^* = AA^{[n-1]} = \bigvee_{k=1}^{n} A^k,$$

and therefore, by (3.44), (3.52), and (3.53), the entries of $\hat{A}$ can be written as

$$\hat{a}_{ij} = \bigvee_{k=1}^{n} a_{ij}^k = \bigvee_{\mu \in T_{ij}^n} l(\mu) = \bigvee_{\mu \in \tilde{T}_{ij}} l(\mu) \quad \text{for all } i, j. \quad (3.56)$$

In words, each element $\hat{a}_{ij}$ is the join of the labels of the non-null elementary paths from x_i to x_j.

Example 3.18. Let P be any path algebra in which the unit element e is the greatest element. (As examples we have the path algebras P_1, P_4, P_5, and P_8.) Then any graph labelled with P is absorptive, since the condition (3.39) is always satisfied. It follows that all matrices in $M_n(P)$ are stable, with stability indices not greater than $n - 1$.

Example 3.19. From the previous example it follows immediately that all Boolean matrices are stable. If A is the Boolean adjacency matrix of a graph G, then A^* has entries $a_{ij}^* = 1$ if there exist any paths from x_i to x_j and $a_{ij}^* = 0$ otherwise; whereas $\hat{A}$ has $\hat{a}_{ij} = 1$ if there exist any non-null paths from x_i to x_j, and $\hat{a}_{ij} = 0$ otherwise.

As a particular case, the Boolean matrix of Fig. 3.3 is stable of index 2 (see Example 3.16), with

$$A^* = A^{[2]} = \begin{bmatrix} 1 & 1 & 0 & 1 \\ 1 & 1 & 0 & 1 \\ 1 & 1 & 1 & 1 \\ 1 & 1 & 0 & 1 \end{bmatrix}, \quad \hat{A} = \begin{bmatrix} 1 & 1 & 0 & 1 \\ 1 & 1 & 0 & 1 \\ 1 & 1 & 0 & 1 \\ 1 & 1 & 0 & 1 \end{bmatrix}.$$

Example 3.20. *Shortest paths.* Let G be a graph labelled with P_2, and let us interpret its arc labels as physical lengths. In this case, G is absorptive if all its cycles are of non-negative length. If G is absorptive then in the closure A^* of the adjacency matrix A of G, each element a_{ij}^* is the distance (i.e. the length of a shortest path) from x_i to x_j.

For the graph of Fig. 3.1 (whose adjacency matrix A is given in Example 3.12, with its powers in Example 3.14),

$$A^* = \bigvee_{k=0}^{3} A^k = \begin{bmatrix} 0 & 2 & \infty & 3 \\ 4 & 0 & \infty & 1 \\ 8 & 4 & 0 & 5 \\ 3 & 5 & \infty & 0 \end{bmatrix}.$$

Finally, we give two useful results relating to absorptive matrices:

(i) *Let A and B be two matrices in $M_n(P)$, such that $B \leqslant A$. Then if A is absorptive, B is also absorptive.*

Let $G_A = (X, U_A)$ and $G_B = (X, U_B)$ be the labelled graphs of A and B respectively. Then since $B \leqslant A$, $U_B \subseteq U_A$, and for any arc $(x_i, x_j) \in U_B$, the label $l_B(x_i, x_j)$ of this arc on G_B is not greater than the label $l_A(x_i, x_j)$ of the corresponding arc on G_A. It follows that since $l(\gamma) \leqslant e$ for every cycle γ on G_A, $l(\gamma) \leqslant e$ for every cycle γ on G_B, which implies that B is absorptive.

(ii) *Let A be any matrix in $M_n(P)$. Then if A is absorptive, A^* is also absorptive.*

Let us assume that A is absorptive. Then by (3.55) all diagonal entries of A^* have the value e, and since $A^*A^* = A^*$, the diagonal elements of all the powers of A^* also have the value e. It follows by (3.44) that for every elementary cycle γ on the graph of A^*, $l(\gamma) \leqslant e$, as required.

3.5. The formulation and solution of path problems

From the examples of the previous section it will be clear that, in algebraic terms, many path problems consist essentially of the determination of one or more elements of the weak or strong closure of an adjacency matrix: Table 3.2 shows how a number of different types of path problems can be formulated in this way.

To obtain the weak or strong closure of a matrix A, it is of course possible to compute successive powers of A and to form their join. However, there are much less laborious methods, which can be used either to form a complete closure matrix, or only particular rows, columns, or other submatrices of it, as required.

TABLE 3.2

Path problem	Path algebra	Significance of arc label $l(x_i, x_j)$	Significance of matrix element a_{ij}^* or $\hat{a}_{ij}$
Existence of paths	P_1	All arcs have the label '1'	$a_{ij}^* = 1$ if x_j is accessible from x_i, and $a_{ij}^* = 0$ otherwise; $\hat{a}_{ij} = 1$ if there exists a non-null path from x_i to x_j, $\hat{a}_{ij} = 0$ otherwise
Shortest paths	P_2	Physical length, or time duration	a_{ij}^* is the length of a shortest path from x_i to x_j
Critical (longest) paths	P_3		a_{ij}^* is the length of a longest path from x_i to x_j
Most reliable paths	P_4	Reliability, i.e. probability of existence of (x_i, x_j)	a_{ij}^* is the reliability of a most reliable path from x_i to x_j
Maximal capacity paths	P_5	Capacity, i.e. maximum rate of flow along (x_i, x_j)	a_{ij}^* is the capacity of a path of maximum capacity from x_i to x_j
Enumeration of all paths	P_6		$\hat{a}_{ij}$ is the set of names of the non-null paths from x_i to x_j
Enumeration of simple paths	P_7	$l(x_i, x_j) = \{n_{ij}\}$, where n_{ij} is the name of (x_i, x_j) —see Example 3.9	$\hat{a}_{ij}$ is the set of names of the non-null simple paths from x_i to x_j
Enumeration of elementary paths	P_8		$\hat{a}_{ij}$ is the set of names of the non-null elementary paths from x_i to x_j

Let us consider the matrix equations

$$Y = AY \vee B \quad \text{and} \quad Y = YA \vee B,$$

where A and B are specified matrices in a path algebra $M_n(P)$. We recall from Section 3.2.4 that (if A is stable) the least solutions of these equations are

$$Y = A^*B \quad \text{and} \quad Y = BA^*$$

respectively, and we observe that with $B = E$ these solutions both become

$$Y = A^*$$

while if $B = A$ they become

$$Y = \hat{A}.$$

Therefore, if we can find a method of solving such equations, it will enable us to compute closure matrices. Furthermore, it will be noted that if we require only the ith column of A^* say, this can be expressed as the least column vector $\mathbf{y}$ which satisfies the simpler equation

$$\mathbf{y} = A\mathbf{y} \vee \mathbf{b} \tag{3.57}$$

where $\mathbf{b}$ is the ith *unit vector*, that is, the ith column vector $\mathbf{e}_i$ of the unit matrix E; similarly, the ith row of A^* can be expressed as the least row vector $\mathbf{y}$ which satisfies the equation

$$\mathbf{y} = \mathbf{y}A \vee \mathbf{b} \tag{3.58}$$

where $\mathbf{b}$ is the ith row of E. (Alternatively, if the multiplicative operation of P is commutative it may be convenient to express the ith row of A^* as the least solution of the equation

$$\mathbf{y} = A'\mathbf{y} \vee \mathbf{b} \tag{3.59}$$

where $\mathbf{b} = \mathbf{e}_{i\cdot}$.) In the same way, the ith column of $\hat{A}$ can be expressed as the least solution of (3.57) with $\mathbf{b} = \mathbf{a}_i$, the ith column of A, and we may also express rows of $\hat{A}$ as solutions of equations of the form (3.58) or (3.59).

Now let us consider the problem of finding the least solution of (3.57), viz.

$$\mathbf{y} = A^*\mathbf{b} = (E \vee A \vee A^2 \vee \cdots)\mathbf{b}. \tag{3.60}$$

In ordinary matrix algebra, we frequently encounter the problem of solving a linear system of the form

$$Cy = b \qquad (3.61)$$

where the matrix C and vector $\mathbf{b}$ are specified. Now by defining the matrix $A = I - C$, where I is the unit matrix, we may write (3.61) as

$$\mathbf{y} = A\mathbf{y} + \mathbf{b}. \qquad (3.62)$$

If the sequence of powers of A converges to a zero matrix, then the series $I + A + A^2 + \cdots$ is convergent, and its sum is equal to $(I - A)^{-1} = C^{-1}$; we may then write the solution of (3.62) as

$$\mathbf{y} = (I - A)^{-1}\mathbf{b} = (I + A + A^2 + \cdots)\mathbf{b}. \qquad (3.63)$$

We observe a certain similarity between our problem (3.57) and the linear algebraic problem (3.62), and between their solutions (3.60) and (3.63). Rather surprisingly, many results in ordinary matrix algebra hold in the algebra $M_n(P)$, if ordinary matrix addition $X + Y$ and multiplication $X \cdot Y$ are interpreted as the operations $X \vee Y$ and XY of $M_n(P)$, and if each inverse $(I - X)^{-1}$ is replaced by a closure X^*.

This similarity suggests the possibility of solving equations in $M_n(P)$ by variants of the classical methods of linear algebra, the new methods differing from the classical ones only in the significance of the elementary operations. In the following sections we shall demonstrate that it is indeed possible to solve systems of the form $\mathbf{y} = A\mathbf{y} \vee \mathbf{b}$ by the classical direct and iterative methods of solving linear systems. Furthermore, we shall find that with different concrete interpretations of the join and multiplicative operations, these methods become well-known algorithms for finding paths on graphs.

3.6. Direct methods of solution

3.6.1. Triangular matrices

Before deriving the general direct methods of solving matrix equations, it will be helpful to consider some properties of triangular matrices.

Let A be a strictly upper or lower triangular matrix belonging to some algebra $M_n(P)$, and let G be its graph. As was shown in

Section 3.4.1, the graph G is acyclic; it follows that A is absorptive, and furthermore, since G does not contain any paths of order greater than $n-1$, A is nilpotent (with $A^n = \phi$).

Now let us consider the problem of solving a system

$$\mathbf{y} = A\mathbf{y} \vee \mathbf{b} \qquad (3.64)$$

where A is a strictly upper or lower triangular matrix, and $\mathbf{b}$ is a column vector. Since A is absorptive, it is stable, so the system has a least solution $\mathbf{y} = A^*\mathbf{b}$. Furthermore, since A is nilpotent, this solution is unique (see Section 3.2.4). The solution can be obtained easily, as follows.

Let us suppose first that A is strictly lower triangular. Then the system (3.64) can be expressed as

$$\left. \begin{aligned} y_1 &= b_1 \\ y_i &= \left(\bigvee_{j=1}^{i-1} a_{ij}y_j \right) \vee b_i \qquad (i = 2, 3, \ldots, n). \end{aligned} \right\} \quad \begin{array}{l} (3.65\text{(a)}) \\[2ex] (3.65\text{(b)}) \end{array}$$

The first equation (3.65(a)) gives y_1 directly. Having found y_1, we may use the equation for y_2 in (3.65(b)) to obtain y_2 directly:

$$y_2 = a_{21}y_1 \vee b_2,$$

and then the equation for y_3 gives this unknown immediately, as

$$y_3 = a_{31}y_1 \vee a_{32}y_2 \vee b_3.$$

Continuing in this fashion, we obtain in turn all the unknowns $y_1, y_2, \ldots, y_n$. This procedure is called the *forward substitution method* of solving (3.65).

If A is strictly *upper* triangular, the system $\mathbf{y} = A\mathbf{y} \vee \mathbf{b}$ can be written as

$$\left. \begin{aligned} y_i &= \left(\bigvee_{j=i+1}^{n} a_{ij}y_j \right) \vee b_i \qquad (i = 1, 2, \ldots, n-1), \\ y_n &= b_n. \end{aligned} \right\} \quad \begin{array}{l} (3.66\text{(a)}) \\[2ex] (3.66\text{(b)}) \end{array}$$

Here, the last equation (3.66(b)) immediately gives y_n. Having found y_n, we can use the equation for y_{n-1} in (3.66(a)) to obtain this unknown, and then we can determine y_{n-2} from the preceding equation, and so on. Continuing in this way, we obtain in turn all the unknowns $y_n, y_{n-1}, y_{n-2}, \ldots, y_1$. This process is called *back-substitution*.

Example 3.21. *Critical path analysis.* In Example 2.11 it was shown how a project can be represented by an 'activity graph'. For the project considered in this example it was easy to determine the lengths of longest paths from the 'start' node to each of the other nodes by inspection, but for large projects this would be impossible.

To find the path lengths systematically, it is convenient to consider an activity graph as a graph labelled with the path algebra P_3, in which case the length of a longest path from a node x_i to a node x_j is given by the element a_{ij}^* of the closure matrix A^*. Thus, if the 'start' node of the activity graph is x_s, the earliest starting time of the ith activity (which is the length of a longest path from x_s to x_i) is given by a_{si}^*. The required starting times are therefore defined by the sth row of A^*—or alternatively the sth column of $(A')^*$. (Since multiplication on P is commutative here, the usual 'reversal rule' applies to transposed products: $(XY)' = Y'X'$, and therefore $(A^*)' = (A')^*$.) Now the sth column of $(A')^*$ is the least solution of the equation

$$\mathbf{y} = A'\mathbf{y} \vee \mathbf{e}_s. \qquad (3.67)$$

If the nodes of the graph are numbered in order of increasing rank then A' is strictly lower triangular, and this equation can be solved by forward substitution.

For the graph of Fig. 2.22 equation (3.67) becomes

$$
\begin{bmatrix} y_1 \\ y_2 \\ y_3 \\ y_4 \\ y_5 \\ y_6 \\ y_7 \\ y_8 \\ y_9 \\ y_{10} \end{bmatrix}
=
\begin{bmatrix}
0 & & & & & & & & & \\
0 & -\infty & & & & & & & & \\
-\infty & 4 & -\infty & & & & -\infty & & & \\
-\infty & 4 & -\infty & -\infty & & & & & & \\
-\infty & 4 & -\infty & -\infty & -\infty & & & & & \\
-\infty & -\infty & -\infty & 6 & 2 & -\infty & & & & \\
-\infty & -\infty & -\infty & -\infty & 2 & -\infty & -\infty & & & \\
-\infty & -\infty & 10 & -\infty & -\infty & 11 & -\infty & 3 & & \\
-\infty & -\infty & -\infty & -\infty & -\infty & -\infty & 22 & -\infty & 17 &
\end{bmatrix}
\begin{bmatrix} y_1 \\ y_2 \\ y_3 \\ y_4 \\ y_5 \\ y_6 \\ y_7 \\ y_8 \\ y_9 \\ y_{10} \end{bmatrix}
\vee
\begin{bmatrix} 0 \\ -\infty \\ -\infty \\ -\infty \\ -\infty \\ -\infty \\ -\infty \\ -\infty \\ -\infty \\ -\infty \end{bmatrix}.
$$

Applying the forward substitution method, with $x \vee y$ and $x \cdot y$ interpreted as $\max\{x, y\}$ and $x + y$ respectively, we obtain successively

$$y_1 = 0,$$

$$y_2 = 0 + y_1 = 0,$$

$$y_3 = 0 + y_1 = 0,$$

$$y_4 = 4 + y_2 = 4,$$

$$y_5 = 4 + y_2 = 4,$$

$$y_6 = 4 + y_2 = 4,$$

$$y_7 = \max\{6 + y_4, 2 + y_5\} = 10,$$

$$y_8 = 2 + y_5 = 6,$$

$$y_9 = \max\{10 + y_3, 11 + y_6, 3 + y_8\} = 15,$$

$$y_{10} = \max\{22 + y_7, 17 + y_9\} = 32.$$

These y-values are the earliest starting times of the activities. The latest starting times can be obtained similarly (by solving $\mathbf{y} = A\mathbf{y} \vee \mathbf{e}_{10}$ by back-substitution).

The techniques described here are essentially those used in practice for critical path analysis of large projects (Leavenworth 1961; Montalbano 1967).

Example 3.22. *Multi-stage decision problems.* The following kind of problem arises for instance in stock control and in planning industrial investments.

At a time t_0, a system is in a known state $s^{(0)}$. At each of r subsequent times t_k, $(k = 1, 2, \ldots, r)$, the system will be in one of several possible states; we denote the number of possible states at time t_k by n_k, and denote these states by $s_1^{(k)}, s_2^{(k)}, \ldots, s_{n_k}^{(k)}$ (see Fig. 3.5). At each time t_k a decision is made, which determines the state of the system at time t_{k+1}. Each feasible transition from a state $s_i^{(k)}$ at time t_k to a state $s_j^{(k+1)}$ at time t_{k+1} has an associated cost $c_{ij}^{(k)}$. The problem is to find an *optimal policy*, i.e. a sequence of decisions at the times $t_0, t_1, \ldots, t_{r-1}$ for which the total expenditure over the period t_0 to t_r is minimal. (For a concrete example, see Exercise 3.4 at the end of the chapter.)

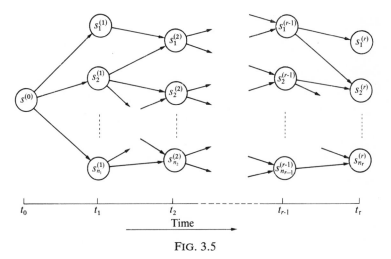

FIG. 3.5

Let us assign a length to each arc on the state diagram of the system, this being the cost of the corresponding state transition. Then a shortest path from the node $s^{(0)}$ to any node $s_i^{(r)}$ representing a state at time t, defines an optimal policy.

To find such a path, we consider the state diagram as a graph labelled with P_2. Then, for a node ordering

$$s^{(0)}, s_1^{(1)}, \ldots, s_{n_1}^{(1)}, s_1^{(2)}, \ldots, s_{n_2}^{(2)}, s_1^{(3)}, \ldots, s_{n_r}^{(r)},$$

the adjacency matrix of the network has the block form

$$A = \begin{bmatrix} D^{(0)} & M^{(0)} & & & \\ & D^{(1)} & M^{(1)} & & \infty \\ & & D^{(2)} & \ddots & \\ & & & \ddots & M^{(r-1)} \\ & \infty & & & D^{(r)} \end{bmatrix},$$

where each diagonal block $D^{(k)}$ is a square matrix of order n_k, whose entries are all ∞, while each block $M^{(k)}$ is a matrix of n_k rows and n_{k+1} columns, with

$$m_{ij}^{(k)} = \begin{cases} c_{ij}^{(k)}, & \text{if the transition from } s_i^{(k)} \text{ to } s_j^{(k+1)} \text{ is feasible,} \\ \infty, & \text{otherwise.} \end{cases}$$

Now let us denote the distance from $s^{(0)}$ to each node $s_i^{(k)}$ by $y_i^{(k)}$. Then the vector of these distances

$$\mathbf{y} = [y_1^{(0)}, y_1^{(1)}, y_2^{(1)}, \ldots, y_{n_r}^{(r)}]'$$

can be expressed as the least solution of the equation

$$\mathbf{y} = A'\mathbf{y} \vee \mathbf{e}_1.$$

Since the matrix A' is lower triangular, this equation can be solved by the forward substitution method. We note that if the vector $\mathbf{y}$ is partitioned into $r+1$ components

$$\mathbf{y} = [\mathbf{y}^{(0)}, \mathbf{y}^{(1)}, \ldots, \mathbf{y}^{(r)}]'$$

where

$$\mathbf{y}^{(k)} = [y_1^{(k)}, y_2^{(k)}, \ldots, y_{n_k}^{(k)}]', \qquad (k = 0, 1, \ldots, r),$$

then the forward substitution method gives each component of $\mathbf{y}$ in turn, as

$$\mathbf{y}^{(0)} = 0, \qquad \mathbf{y}^{(k)} = (M^{(k-1)})'\mathbf{y}^{(k-1)}, \qquad (k = 1, 2, \ldots, r).$$

This algorithm is well-known, as a form of *dynamic programming* (Bellman 1957; White 1969).

3.6.2. The Gauss elimination method†

Let us now consider the problem of finding the least solution of a system $\mathbf{y} = A\mathbf{y} \vee \mathbf{b}$, where A is any stable $n \times n$ matrix.

To define the Gauss method, we denote the system to be solved by

$$\mathbf{y} = A^{(0)}\mathbf{y} \vee \mathbf{b}^{(0)}. \tag{3.68}$$

From this system we shall derive n new systems

$$\mathbf{y} = A^{(k)}\mathbf{y} \vee \mathbf{b}^{(k)}, \qquad (k = 1, 2, \ldots, n) \tag{3.69}$$

which are all equivalent to the original one in that they all have the same least solution, and where the matrix $A^{(n)}$ of the final system is strictly upper triangular. (The final system can then be solved by back-substitution.)

Essentially, we obtain the kth system from the previous one by using its equation for y_k to eliminate this unknown from the right-hand sides of the equations for $y_k, y_{k+1}, \ldots, y_n$. To define the process formally, let us write the matrix $A^{(k-1)}$ and vector $\mathbf{b}^{(k-1)}$ of the $(k-1)$th system in partitioned form, as

$$A^{(k-1)} = \begin{bmatrix} A_{11}^{(k-1)} & A_{12}^{(k-1)} & A_{13}^{(k-1)} \\ A_{21}^{(k-1)} & A_{22}^{(k-1)} & A_{23}^{(k-1)} \\ A_{31}^{(k-1)} & A_{32}^{(k-1)} & A_{33}^{(k-1)} \end{bmatrix}, \qquad \mathbf{b}^{(k-1)} = \begin{bmatrix} \mathbf{b}_1^{(k-1)} \\ \mathbf{b}_2^{(k-1)} \\ \mathbf{b}_3^{(k-1)} \end{bmatrix},$$

$$(k = 1, 2, \ldots, n), \tag{3.70}$$

where the diagonal submatrices $A_{11}^{(k-1)}$, $A_{22}^{(k-1)}$, and $A_{33}^{(k-1)}$ are square, of order $k-1$, 1, and $n-k$ respectively. (Note that in $A^{(0)}$, the first row and column of this partition do not exist.) Using the same partitioning as (3.70), we define two matrices $Q^{(k)}$ and $R^{(k)}$ by

$$Q^{(k)} = \begin{bmatrix} \Phi & \Phi & \Phi \\ \Phi & A_{22}^{(k-1)} & \Phi \\ \Phi & A_{32}^{(k-1)} & \Phi \end{bmatrix}, \qquad R^{(k)} = \begin{bmatrix} A_{11}^{(k-1)} & A_{12}^{(k-1)} & A_{13}^{(k-1)} \\ A_{21}^{(k-1)} & \Phi & A_{23}^{(k-1)} \\ A_{31}^{(k-1)} & \Phi & A_{33}^{(k-1)} \end{bmatrix},$$

$$(k = 1, 2, \ldots, n). \tag{3.71}$$

---

† To be precise, the method presented here is the counterpart in $M_n(P)$ of Crout's variant of the Gauss elimination method (Fox 1964).

Then (assuming that each $Q^{(k)}$ matrix is stable) the successive matrices $A^{(k)}$ and vectors $\mathbf{b}^{(k)}$ in (3.69) are constructed using

$$\left.\begin{aligned} A^{(k)} &= Q^{(k)*}R^{(k)} \\ \mathbf{b}^{(k)} &= Q^{(k)*}\mathbf{b}^{(k-1)} \end{aligned}\right\} \qquad (k = 1, 2, \ldots, n). \qquad (3.72)$$

To prove that all the systems (3.69) obtained in this way have the same least solution as (3.68), we note from (3.70) and (3.71) that

$$A^{(k-1)} = Q^{(k)} \vee R^{(k)} \qquad (k = 1, 2, \ldots, n). \qquad (3.73)$$

It follows by (3.72), (3.27), and (3.73) that (on condition that $A^{(k)}$ is stable),

$$\begin{aligned} A^{(k)*}\mathbf{b}^{(k)} &= (Q^{(k)*}R^{(k)})^{*}Q^{(k)*}\mathbf{b}^{(k-1)} \\ &= (Q^{(k)} \vee R^{(k)})^{*}\mathbf{b}^{(k-1)} = A^{(k-1)*}\mathbf{b}^{(k-1)}, \\ &\qquad\qquad (k = 1, 2, \ldots, n), \qquad (3.74) \end{aligned}$$

as required.

It is also easy to show that in each matrix $A^{(k)}$, all elements on and below the principal diagonal, in the first k columns, are null: indeed, this condition obviously holds for $A^{(0)}$; so let us assume that it holds for some $A^{(k-1)}$, where $1 \le k < n$, and show that it holds for $A^{(k)}$. From (3.71),

$$(Q^{(k)})^{s} = \begin{bmatrix} \Phi & \Phi & \Phi \\ \Phi & (A_{22}^{(k-1)})^{s} & \Phi \\ \Phi & A_{32}^{(k-1)}(A_{22}^{(k-1)})^{s-1} & \Phi \end{bmatrix}, \qquad (s = 1, 2, \ldots),$$

and therefore

$$Q^{(k)*} = \begin{bmatrix} E & \Phi & \Phi \\ \Phi & A_{22}^{(k-1)*} & \Phi \\ \Phi & A_{32}^{(k-1)}A_{22}^{(k-1)*} & E \end{bmatrix}. \qquad (3.75)$$

Also, since $A_{21}^{(k)}$ and $A_{31}^{(k)}$ are both null (by assumption), we may write (cf. (3.71))

$$R^{(k)} = \begin{bmatrix} A_{11}^{(k-1)} & A_{12}^{(k-1)} & A_{13}^{(k-1)} \\ \Phi & \Phi & A_{23}^{(k-1)} \\ \Phi & \Phi & A_{33}^{(k-1)} \end{bmatrix}. \qquad (3.76)$$

By (3.72), (3.75), and (3.76)

$$A^{(k)} = Q^{(k)} * R^{(k)} = \begin{bmatrix} A_{11}^{(k-1)} & A_{12}^{(k-1)} & A_{13}^{(k-1)} \\ \Phi & \Phi & A_{22}^{(k-1)} * A_{23}^{(k-1)} \\ \Phi & \Phi & A_{33}^{(k-1)} \vee A_{32}^{(k-1)} A_{22}^{(k-1)} * A_{23}^{(k-1)} \end{bmatrix}.$$

$$(3.77)$$

Since $A_{11}^{(k-1)}$ is strictly upper triangular (by assumption), all the elements of $A^{(k)}$ which lie on or below the diagonal in the first k columns are null, as required. Thus, in the final system, the matrix $A^{(n)}$ will be strictly upper triangular.

With regard to the calculation of the successive $A^{(k)}$ matrices, it follows from (3.77) that their entries can be calculated using

$$a_{ij}^{(k)} = \begin{cases} (a_{kk}^{(k-1)})^* a_{kj}^{(k-1)} & \text{if } i = k, j > k, \\ \phi & \text{if } i \geq k, j = k, \\ a_{ij}^{(k-1)} \vee a_{ik}^{(k-1)} (a_{kk}^{(k-1)})^* a_{kj}^{(k-1)} & \text{if } i > k, j > k, \\ a_{ij}^{(k-1)} & \text{otherwise.} \end{cases} \quad (3.78)$$

(In practice, it is convenient to record these computations simply by making successive modifications to the original matrix $A^{(0)}$. When this is done, it is of course not necessary to nullify sub-diagonal entries at any stage.)

To construct the successive $\mathbf{b}^{(k)}$ vectors, we have from (3.72) and (3.75) that

$$\mathbf{b}^{(k)} = Q^{(k)} * \mathbf{b}^{(k-1)} = \begin{bmatrix} \mathbf{b}_1^{(k-1)} \\ A_{22}^{(k-1)} * \mathbf{b}_2^{(k-1)} \\ \mathbf{b}_3^{(k-1)} \vee A_{32}^{(k-1)} A_{22}^{(k-1)} * \mathbf{b}_2^{(k-1)} \end{bmatrix}, \quad (3.79)$$

so the elements of $\mathbf{b}^{(k)}$ are given by

$$b_i^{(k)} = \begin{cases} b_i^{(k-1)} & \text{if } i < k, \\ (a_{kk}^{(k-1)})^* b_k^{(k-1)} & \text{if } i = k, \\ b_i^{(k-1)} \vee a_{ik}^{(k-1)} (a_{kk}^{(k-1)})^* b_k^{(k-1)} & \text{if } i > k. \end{cases}$$

$$(k = 1, 2, \ldots, n). \quad (3.80)$$

If solutions are required for more than one $\mathbf{b}$-vector, but the same matrix A, we can of course perform the transformations of A as before, and apply the transformations (3.79) to all the $\mathbf{b}$-vectors together. A separate back-substitution must then be carried out for each vector.

Gauss elimination for absorptive matrices. In deriving the Gauss elimination method it was necessary to assume that the matrices $Q^{(k)}$ and $A^{(k)}$ are stable at every stage (see (3.72) and (3.74)). It can be demonstrated that these conditions hold if the initial matrix $A^{(0)}$ is absorptive: for if $A^{(0)}$ is absorptive then, since $Q^{(1)} \leqslant A^{(0)}$, the matrix $Q^{(1)}$ is also absorptive (and therefore stable). In this case the matrix $A^{(1)}$ is well defined by (3.72), and since $Q^{(1)} \leqslant A^{(0)}$ and $R^{(1)} \leqslant A^{(0)}$ it follows by (3.72) that

$$A^{(1)} = Q^{(1)*} R^{(1)} \leqslant A^{(0)*} A^{(0)} \leqslant A^{(0)*}$$

which implies that $A^{(1)}$ is absorptive. By repetition of this argument it follows that the successive $A^{(k)}$ and $Q^{(k)}$ matrices are all absorptive, and consequently stable.

We observe that, since the diagonal elements of an absorptive matrix are necessarily sub-unitary, the 'pivotal' element $a_{kk}^{(k-1)}$ in (3.78) and (3.80) has a closure $(a_{kk}^{(k-1)})^* = e$. Thus when $A^{(0)}$ is absorptive these formulae can be simplified to

$$a_{ij}^{(k)} = \begin{cases} \phi & \text{if } i \geq k, j = k, \\ a_{ij}^{(k-1)} \vee a_{ik}^{(k-1)} a_{kj}^{(k-1)} & \text{if } i > k, j > k, \\ a_{ij}^{(k-1)} & \text{otherwise,} \end{cases} \tag{3.81}$$

and

$$b_i^{(k)} = \begin{cases} b_i^{(k-1)} & \text{if } i \leq k, \\ b_i^{(k-1)} \vee a_{ik}^{(k-1)} b_k^{(k-1)} & \text{if } i > k. \end{cases} \tag{3.82}$$

respectively. Furthermore, it is evident from (3.82) that $\mathbf{b}^{(n)} = \mathbf{b}^{(n-1)}$ so it is only necessary to perform $n-1$ transformations of the **b**-vector.

Example 3.23. *Shortest path calculations.* Let us consider the problem of finding the distances from each of the nodes to the node '4' on the graph of Fig. 3.6.

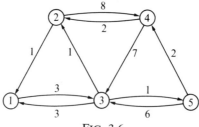

FIG. 3.6

We denote the distance from each node x_i to x_4 by y_i. Then, using P_2, the vector $\mathbf{y} = [y_1, y_2, y_3, y_4, y_5]'$ of these distances can be obtained as the least solution of the equation $\mathbf{y} = A\mathbf{y} \vee \mathbf{b}$, where

$$
A = \begin{bmatrix}
\infty & \infty & 3 & \infty & \infty \\
1 & \infty & \infty & 8 & \infty \\
3 & 1 & \infty & \infty & 1 \\
\infty & 2 & 7 & \infty & \infty \\
\infty & \infty & 6 & 2 & \infty
\end{bmatrix}, \quad \text{and} \quad \mathbf{b} = \mathbf{e}_4 = \begin{bmatrix}
\infty \\
\infty \\
\infty \\
0 \\
\infty
\end{bmatrix}.
$$

Since the graph does not contain any cycles of negative length the matrix A is absorptive, and therefore we can use the simplified form (3.81), (3.82) of the Gauss elimination method. The successive matrices $A^{(k)}$ and vectors $\mathbf{b}^{(k)}$ obtained by this algorithm are given below. (The shading indicates nullified elements of the lower triangle, and the circles indicate other elements whose values change at each stage.)

$$
A^{(1)} = \begin{bmatrix}
\infty & \infty & 3 & \infty & \infty \\
1 & \infty & ④ & 8 & \infty \\
3 & 1 & ⑥ & \infty & 1 \\
\infty & 2 & 7 & \infty & \infty \\
\infty & \infty & 6 & 2 & \infty
\end{bmatrix}, \quad \mathbf{b}^{(1)} = \begin{bmatrix}
\infty \\
\infty \\
\infty \\
0 \\
\infty
\end{bmatrix};
$$

$$
A^{(2)} = \begin{bmatrix}
\infty & \infty & 3 & \infty & \infty \\
1 & \infty & 4 & 8 & \infty \\
3 & 1 & ⑤ & ⑨ & 1 \\
\infty & 2 & ⑥ & ⑩ & \infty \\
\infty & \infty & 6 & 2 & \infty
\end{bmatrix}, \quad \mathbf{b}^{(2)} = \begin{bmatrix}
\infty \\
\infty \\
\infty \\
0 \\
\infty
\end{bmatrix};
$$

$$
A^{(3)} = \begin{bmatrix}
\infty & \infty & 3 & \infty & \infty \\
1 & \infty & 4 & 8 & \infty \\
3 & 1 & 5 & 9 & 1 \\
\infty & 2 & 6 & 10 & ⑦ \\
\infty & \infty & 6 & 2 & ⑦
\end{bmatrix}, \quad \mathbf{b}^{(3)} = \begin{bmatrix}
\infty \\
\infty \\
\infty \\
0 \\
\infty
\end{bmatrix};
$$

$$
A^{(4)} = \begin{bmatrix}
\infty & \infty & 3 & \infty & \infty \\
1 & \infty & 4 & 8 & \infty \\
3 & 1 & 5 & 9 & 1 \\
\infty & 2 & 6 & 10 & 7 \\
\infty & \infty & 6 & 2 & 7
\end{bmatrix}, \quad \mathbf{b}^{(4)} = \begin{bmatrix}
\infty \\
\infty \\
\infty \\
0 \\
②
\end{bmatrix};
$$

$$A^{(5)} = \begin{bmatrix} \infty & \infty & 3 & \infty & \infty \\ 1 & \infty & 4 & 8 & \infty \\ 3 & 1 & 5 & 9 & 1 \\ \infty & 2 & 6 & 10 & 7 \\ \infty & \infty & 6 & 2 & 7 \end{bmatrix}, \qquad \mathbf{b}^{(5)} = \begin{bmatrix} \infty \\ \infty \\ \infty \\ 0 \\ 2 \end{bmatrix}.$$

By applying the back-substitution method of Section 3.6.1 to the system $\mathbf{y} = A^{(5)}\mathbf{y} \vee \mathbf{b}^{(5)}$ we obtain the solution

$$\mathbf{y} = [6 \ 7 \ 3 \ 0 \ 2]'.$$

If all the cycles on a graph are of strictly positive length then all shortest paths are elementary. In this case it is possible to determine the sequence of nodes traversed by a shortest path, from node x_p to node x_q say, in the following manner. First, we construct the vector $\mathbf{y}$ of distances from each node to node x_q, as above. If $y_p = \infty$, there are no paths from x_p to x_q; otherwise, we execute the following algorithm:

Step 1   Let $s_0 = p$ and let $k = 1$.

Step 2   Let s_k be any index such that $a_{s_{k-1}s_k} + y_{s_k} = y_{s_{k-1}}$.

Step 3   If $s_k = q$ then halt; otherwise record s_k, increase k by 1, and return to *Step 2*.

It is easily verified that the sequence of indices $s_1, s_2, \ldots, s_{k-1}$ produced by this algorithm defines a shortest path from x_p to x_q:

$$x_p \rightarrow x_{s_1} \rightarrow x_{s_2} \rightarrow \cdots \rightarrow x_q.$$

As an illustration, to find a shortest path from x_1 to x_4 on the graph given above, we first set $s_0 = 1$; we then set

$$s_1 = 3 \ (\text{since } a_{13} + y_3 = y_1);$$
$$s_2 = 5 \ (\text{since } a_{35} + y_5 = y_3);$$
$$s_3 = 4 \ (\text{since } a_{54} + y_4 = y_5).$$

These indices define the shortest path $x_1 \rightarrow x_3 \rightarrow x_5 \rightarrow x_4$.

Example 3.24. *Determination of simple paths.* Let us consider the problem of finding the simple paths on the graph of Fig. 3.2, from each of the nodes to the node '2'. Using the algebra P_7, the required paths are given by the second column of $\hat{A}$ (see Example 3.17), which can be expressed as the (unique) solution y of the system $\mathbf{y} = A\mathbf{y} \vee \mathbf{a}_2$, where $\mathbf{a}_2$ is the second column of A.

In this system the A-matrix is not absorptive, but it is easily demonstrated that all matrices on P_7 are stable (the proof is left as an exercise for the reader). Applying the Gauss elimination method (3.78), (3.80) we obtain:

$$A^{(0)} = \begin{bmatrix} \phi & \{a\} & \phi & \phi \\ \{b\} & \phi & \phi & \{c\} \\ \phi & \{d\} & \phi & \{e\} \\ \{f\} & \phi & \phi & \{g\} \end{bmatrix}, \qquad \mathbf{b}^{(0)} = \begin{bmatrix} \{a\} \\ \phi \\ \{d\} \\ \phi \end{bmatrix};$$

$$A^{(1)} = \begin{bmatrix} \phi & \{a\} & \phi & \phi \\ \phi & \{ba\} & \phi & \{c\} \\ \phi & \{d\} & \phi & \{e\} \\ \phi & \{fa\} & \phi & \{g\} \end{bmatrix}, \qquad \mathbf{b}^{(1)} = \begin{bmatrix} \{a\} \\ \{ba\} \\ \{d\} \\ \{fa\} \end{bmatrix};$$

$$A^{(2)} = \begin{bmatrix} \phi & \{a\} & \phi & \phi \\ \phi & \phi & \phi & \{c, bac\} \\ \phi & \phi & \phi & \{e, dc, dbac\} \\ \phi & \phi & \phi & \{g, fac\} \end{bmatrix}, \qquad \mathbf{b}^{(2)} = \begin{bmatrix} \{a\} \\ \{ba\} \\ \{d, dba\} \\ \{fa\} \end{bmatrix};$$

$$A^{(3)} = \begin{bmatrix} \phi & \{a\} & \phi & \phi \\ \phi & \phi & \phi & \{c, bac\} \\ \phi & \phi & \phi & \{e, dc, dbac\} \\ \phi & \phi & \phi & \{g, fac\} \end{bmatrix}, \qquad \mathbf{b}^{(3)} = \begin{bmatrix} \{a\} \\ \{ba\} \\ \{d, dba\} \\ \{fa\} \end{bmatrix};$$

$$A^{(4)} = \begin{bmatrix} \phi & \{a\} & \phi & \phi \\ \phi & \phi & \phi & \{c, bac\} \\ \phi & \phi & \phi & \{e, dc, dbac\} \\ \phi & \phi & \phi & \phi \end{bmatrix}, \qquad \mathbf{b}^{(4)} = \begin{bmatrix} \{a\} \\ \{ba\} \\ \{d, dba\} \\ \{fa, gfa\} \end{bmatrix}.$$

Then, application of the back-substitution method to the system $\mathbf{y} = A^{(4)}\mathbf{y} \vee \mathbf{b}^{(4)}$ gives

$$\mathbf{y} = \begin{bmatrix} \{a\} \\ \{cfa, cgfa, ba\} \\ \{efa, egfa, dcfa, dcgfa, d, dba\} \\ \{fa, gfa\} \end{bmatrix}.$$

The same technique can be used to determine the *elementary* paths and cycles of a graph: if each arc is labelled with the index of its terminal endpoint, viz. $l(x_i, x_j) = \{j\}$, then each element $\hat{a}_{ij}$ of $\hat{A}$ gives the set of node sequences of the elementary paths from x_i to x_j.

3.6.3. The Jordan elimination method

As an alternative to the Gauss method, it is possible to perform the n successive transformations of the system $\mathbf{y} = A^{(0)}\mathbf{y} \vee \mathbf{b}^{(0)}$ in such a way that in the final system the matrix $A^{(n)}$ is null, so that no back-substitutions are required. Essentially, the difference between the transformations performed in the Jordan method and those of the Gauss method is that, in the kth step of the Jordan method, we use the kth equation to eliminate y_k from all the other equations of the system, and not just the equations for $y_k, y_{k+1}, \ldots, y_n$. To achieve this, we simply replace the matrices $Q^{(k)}$ and $R^{(k)}$ of (3.72) by

$$Q^{(k)} = \begin{bmatrix} \Phi & A_{12}^{(k-1)} & \Phi \\ \Phi & A_{22}^{(k-1)} & \Phi \\ \Phi & A_{32}^{(k-1)} & \Phi \end{bmatrix}, \qquad R^{(k)} = \begin{bmatrix} A_{11}^{(k-1)} & \Phi & A_{13}^{(k-1)} \\ A_{21}^{(k-1)} & \Phi & A_{23}^{(k-1)} \\ A_{31}^{(k-1)} & \Phi & A_{33}^{(k-1)} \end{bmatrix},$$

$$(k = 1, 2, \ldots, n). \qquad (3.83)$$

It is evident from (3.83) that, as in the Gauss method,

$$A^{(k-1)} = Q^{(k)} \vee R^{(k)}, \qquad (k = 1, 2, \ldots, n),$$

which implies (through (3.74)) that all the systems $\mathbf{y} = A^{(k)}\mathbf{y} \vee \mathbf{b}^{(k)}$ have the same least solution.

Next, we can show by induction on k that in each matrix $A^{(k)}$, the first k columns are null: indeed, it is evident from (3.83) that if the first $k-1$ columns of $A^{(k-1)}$ are null, then the first k columns of $R^{(k)}$ are null, which implies by (3.72) that the first k columns of $A^{(k)}$ are null, as required.

Thus in the final system $\mathbf{y} = A^{(n)}\mathbf{y} \vee \mathbf{b}^{(n)}$ the matrix $A^{(n)}$ is null, so this system gives the required solution directly, as $\mathbf{y} = \mathbf{b}^{(n)}$.

With regard to the calculation of the successive matrices $A^{(k)}$ and vectors $\mathbf{b}^{(k)}$, it is evident from (3.83) that

$$Q^{(k)*} = \begin{bmatrix} E & A_{12}^{(k-1)}A_{22}^{(k-1)*} & \Phi \\ \Phi & A_{22}^{(k-1)*} & \Phi \\ \Phi & A_{32}^{(k-1)}A_{22}^{(k-1)*} & E \end{bmatrix}, \qquad (k = 1, 2, \ldots, n)$$

$$(3.84)$$

and, since the first k columns of $R^{(k)}$ in (3.83) are null, we can write this matrix as

$$R^{(k)} = \begin{bmatrix} \Phi & \Phi & A_{13}^{(k-1)} \\ \Phi & \Phi & A_{23}^{(k-1)} \\ \Phi & \Phi & A_{33}^{(k-1)} \end{bmatrix}, \qquad (k = 1, 2, \ldots, n). \qquad (3.85)$$

It follows from (3.72) that

$$A^{(k)} = \begin{bmatrix} \Phi & \Phi & A_{13}^{(k-1)} \vee A_{12}^{(k-1)} A_{22}^{(k-1)*} A_{23}^{(k-1)} \\ \Phi & \Phi & A_{22}^{(k-1)*} A_{23}^{(k-1)} \\ \Phi & \Phi & A_{33}^{(k-1)} \vee A_{32}^{(k-1)} A_{22}^{(k-1)*} A_{23}^{(k-1)} \end{bmatrix}$$

and

$$\mathbf{b}^{(k)} = \begin{bmatrix} \mathbf{b}_1^{(k-1)} \vee A_{12}^{(k-1)} A_{22}^{(k-1)*} \mathbf{b}_2^{(k-1)} \\ A_{22}^{(k-1)*} \mathbf{b}_2^{(k-1)} \\ \mathbf{b}_3^{(k-1)} \vee A_{32}^{(k-1)} A_{22}^{(k-1)*} \mathbf{b}_2^{(k-1)} \end{bmatrix} \qquad (k = 1, 2, \ldots, n). \tag{3.86}$$

As in the derivation of the Gauss method, it has been assumed here that the matrices $Q^{(k)}$ and $A^{(k)}$ are stable at every stage. In the particular case where $A^{(0)}$ is absorptive, this condition holds (the proof given for the Gauss method being valid here also). Again, if $A^{(0)}$ is absorptive the formulae (3.86) can be simplified, by setting $A_{22}^{(k-1)*} = e$ throughout.

The computation of closure matrices. If we require the least solution of a system $Y = AY \vee B$ where B is an $n \times m$ matrix, we can of course obtain this by performing the transformations of A as before, and operating on each column of B in the same way as we treated the single $\mathbf{b}$-vector previously. Thus, writing $A^{(0)} = A$ and $B^{(0)} = B$, we compute successively (cf. (3.72))

$$\left. \begin{array}{l} A^{(k)} = Q^{(k)*} R^{(k)} \\ B^{(k)} = Q^{(k)*} B^{(k-1)} \end{array} \right\} \qquad (k = 1, 2, \ldots, n), \tag{3.87}$$

and on termination we have $B^{(n)} = A^*B$.

In particular, this technique can be used to compute the weak closure of a matrix A: since $\hat{A}$ is the least solution of the equation $Y = AY \vee A$, if we set $A^{(0)} = B^{(0)} = A$ and then apply (3.87), this gives $B^{(n)} = \hat{A}$. Furthermore, since $A^{(0)} = B^{(0)}$ initially, it follows from (3.83) and (3.87) that at every stage, the non-null columns of $A^{(k)}$ are identical to the corresponding columns of $B^{(k)}$, and therefore in place of (3.84) we may write

$$Q^{(k)*} = \begin{bmatrix} E & B_{12}^{(k-1)} B_{22}^{(k-1)*} & \Phi \\ \Phi & B_{22}^{(k-1)*} & \Phi \\ \Phi & B_{32}^{(k-1)} B_{22}^{(k-1)*} & E \end{bmatrix}, \qquad (k = 1, 2, \ldots, n). \tag{3.88}$$

Thus, by (3.87) and (3.88), each matrix $B^{(k)}$ can be defined directly in terms of $B^{(k-1)}$:

$$B^{(k)} = Q^{(k)} * B^{(k-1)}$$

$$= \begin{bmatrix} B_{11}^{(k-1)} \vee B_{12}^{(k-1)} B_{22}^{(k-1)} * B_{21}^{(k-1)} & B_{12}^{(k-1)} B_{22}^{(k-1)} * & B_{13}^{(k-1)} \vee B_{12}^{(k-1)} B_{22}^{(k-1)} * B_{23}^{(k-1)} \\ B_{22}^{(k-1)} * B_{21}^{(k-1)} & B_{22}^{(k-1)} * B_{22}^{(k-1)} & B_{22}^{(k-1)} * B_{23}^{(k-1)} \\ B_{31}^{(k-1)} \vee B_{32}^{(k-1)} B_{22}^{(k-1)} * B_{21}^{(k-1)} & B_{32}^{(k-1)} B_{22}^{(k-1)} * & B_{33}^{(k-1)} \vee B_{32}^{(k-1)} B_{22}^{(k-1)} * B_{23}^{(k-1)} \end{bmatrix}$$

$$(k = 1, 2, \ldots, n), \qquad (3.89)$$

and therefore its elements can be computed using successively

$$b_{ij}^{(k)} = \begin{cases} b_{ik}^{(k-1)} (b_{kk}^{(k-1)})^* & \text{if } j = k, \\ (b_{kk}^{(k-1)})^* b_{kj}^{(k-1)} & \text{if } i = k, \\ b_{ij}^{(k-1)} \vee b_{ik}^{(k-1)} (b_{kk}^{(k-1)})^* b_{kj}^{(k-1)} & \text{if } i, j \neq k, \end{cases}$$

$$(k = 1, 2, \ldots, n). \qquad (3.90)$$

In the particular case where A is absorptive this formula simplifies to

$$b_{ij}^{(k)} = \begin{cases} b_{ij}^{(k-1)} & \text{if } i = k \text{ or } j = k, \\ b_{ij}^{(k-1)} \vee b_{ik}^{(k-1)} b_{kj}^{(k-1)} & \text{if } i, j \neq k, \end{cases}$$

$$(k = 1, 2, \ldots, n). \qquad (3.91)$$

The algorithm (3.90) has a counterpart in linear algebra—the well-known Jordan method of matrix inversion. It has been invented for regular languages by McNaughton and Yamada (1960). The simplified algorithm (3.91) has also been invented as a method of finding closures of Boolean matrices by Roy (1959) and Warshall (1962). In the context of shortest path problems it is known as Floyd's algorithm (Floyd 1962), and it has been invented as a method of listing elementary paths by Murchland (1965).

Example 3.25. *Determination of elementary paths and cycles.* Let G be a graph whose arcs have names—as in Fig. 3.2—and let us consider these arc labels as elements of the path algebra P_8. Then for any path μ on the graph, $l(\mu) = \{\sigma\}$ where σ is the name of μ (as defined in Example 3.9).

Let A be the adjacency matrix of G. Now all matrices in $M_n(P_8)$ are absorptive (see Example 3.18), and by (3.56), and the definition of the join operation (given in Section 3.2.2), the matrix $\hat{A}$ has entries

$$\hat{a}_{ij} = \bigvee_{\mu \in \hat{T}_{ij}} l(\mu) = b\left(\bigvee_{\mu \in \hat{T}_{ij}} l(\mu)\right) = b(N_{ij})$$

where N_{ij} is the set of names of the elementary paths from x_i to x_j. Now if μ is an elementary path from x_i to x_j, no proper subset of the arc set of μ forms a path from x_i to x_j; it follows that each path name in N_{ij} is basic to N_{ij}, and therefore

$$\hat{a}_{ij} = b(N_{ij}) = N_{ij}.$$

The matrix $\hat{A}$ can be computed using the simplified form (3.91) of the Jordan 'inversion' method: for the graph of Fig. 3.2,

$$B^{(0)} = A = \begin{bmatrix} \phi & \{a\} & \phi & \phi \\ \{b\} & \phi & \phi & \{c\} \\ \phi & \{d\} & \phi & \{e\} \\ \{f\} & \phi & \phi & \{g\} \end{bmatrix}; \quad B^{(1)} = \begin{bmatrix} \phi & \{a\} & \phi & \phi \\ \{b\} & \{ba\} & \phi & \{c\} \\ \phi & \{d\} & \phi & \{e\} \\ \{f\} & \{fa\} & \phi & \{g\} \end{bmatrix};$$

$$B^{(2)} = \begin{bmatrix} \{ab\} & \{a\} & \phi & \{ac\} \\ \{b\} & \{ba\} & \phi & \{c\} \\ \{db\} & \{d\} & \phi & \{e, dc\} \\ \{f\} & \{fa\} & \phi & \{g, fac\} \end{bmatrix}; \quad B^{(3)} = B^{(2)};$$

$$B^{(4)} = \begin{bmatrix} \{ab, acf\} & \{a\} & \phi & \{ac\} \\ \{b, cf\} & \{ba, cfa\} & \phi & \{c\} \\ \{db, ef, dcf\} & \{d, efa\} & \phi & \{e, dc\} \\ \{f\} & \{fa\} & \phi & \{g, fac\} \end{bmatrix}.$$

3.7. Iterative methods

3.7.1. The Jacobi, Gauss–Seidel, and double-sweep iterative methods

In this section we shall present three iterative methods for finding the least solution $\mathbf{y} = A^*\mathbf{b}$ of a system

$$\mathbf{y} = A\mathbf{y} \vee \mathbf{b}, \tag{3.92}$$

where the matrix A is stable. For simplicity, it will be assumed throughout that the diagonal entries of A are all null. We note that, should an A-matrix not satisfy this condition, we may express it as the join of two matrices:

$$A = C \vee D, \tag{3.93}$$

where C is the matrix obtained from A by nullifying its diagonal elements, and $D = diag\,(a_{11}, a_{22}, \ldots, a_{nn})$. It follows by (3.27) that (on condition that all the diagonal elements a_{ii} are stable) the least solution of (3.92) can be written as

$$A^*\mathbf{b} = (C \vee D)^*\mathbf{b} = (D^*C)^*D^*\mathbf{b},$$

and it can therefore be obtained as the least solution of the system

$$\mathbf{y} = \tilde{A}\mathbf{y} \vee \tilde{\mathbf{b}} \tag{3.94(a)}$$

where

$$\tilde{A} = D^*C \quad \text{and} \quad \tilde{\mathbf{b}} = D^*\mathbf{b}. \tag{3.94(b)}$$

From (3.94(b)) the entries of the matrix $\tilde{A} = [\tilde{a}_{ij}]$ and the vector $\tilde{\mathbf{b}} = [\tilde{b}_i]$ can be written as

$$\tilde{a}_{ij} = \begin{cases} (a_{ii})^*a_{ij} & \text{if } i \neq j, \\ \phi & \text{if } i = j, \end{cases} \quad \text{and} \quad \tilde{b}_i = (a_{ii})^*b_i;$$

thus, in the new system (3.94), all diagonal entries of the matrix $\tilde{A}$ are null.

In an iterative method of solving the system (3.92), we first choose some estimate $\mathbf{y}^{(0)}$ of the required solution, and then we derive from it a sequence of successive estimates $\mathbf{y}^{(k)}$, $(k = 1, 2, \ldots)$; if the method is successful, these vectors ultimately take the value of the required solution. Some techniques for constructing the $\mathbf{y}^{(k)}$ vectors are defined below.

(i) *The Jacobi method.* The form of the system (3.92) immediately suggests the iterative scheme

$$\mathbf{y}^{(k)} = A\mathbf{y}^{(k-1)} \vee \mathbf{b}, \qquad (k = 1, 2, \ldots). \tag{3.95}$$

This iterative method is the counterpart of the Jacobi method of solving a system $\mathbf{y} = A\mathbf{y} + \mathbf{b}$ in linear algebra (Varga 1962). The concrete interpretation of (3.95) for the path algebra P_2 is also widely known, as *Bellman's method* of solving shortest-path problems (Bellman 1958).

It will be noted that, since all diagonal elements of A are null, the formula for calculating the elements of the kth estimate $\mathbf{y}^{(k)}$ can be written as

$$y_i^{(k)} = \left(\bigvee_{\substack{j=1 \\ j \neq i}}^{n} a_{ij} y_j^{(k-1)} \right) \vee b_i, \qquad (i = 1, 2, \ldots, n). \tag{3.96}$$

(ii) *The Gauss–Seidel method.* From the formula (3.96) it is clear that in the Jacobi method one only uses elements of $\mathbf{y}^{(k-1)}$ in calculating $\mathbf{y}^{(k)}$, but intuitively it would seem reasonable to use always the latest available estimates of the components $\tilde{y}_i$ of the required solution $\tilde{\mathbf{y}}$. This leads to the iterative method in which each of the elements $y_1^{(k)}, y_2^{(k)}, \ldots, y_n^{(k)}$ are obtained in turn using (cf. (3.96))

$$y_i^{(k)} = \left(\bigvee_{j=1}^{i-1} a_{ij} y_j^{(k)} \right) \vee \left(\bigvee_{j=i+1}^{n} a_{ij} y_j^{(k-1)} \right) \vee b_i,$$

$$(i = 1, 2, \ldots, n). \tag{3.97}$$

This method has the additional advantage over the Jacobi method that it does not require the simultaneous storage of the two approximations $y_i^{(k-1)}$ and $y_i^{(k)}$ in the course of computation.

This procedure can be defined in matrix form, by writing

$$A = L \vee U \tag{3.98}$$

where L and U are respectively strictly lower and strictly upper triangular matrices, whose non-null entries are the entries of A respectively below and above its main diagonal. Then in matrix notation, (3.97) becomes

$$\mathbf{y}^{(k)} = L\mathbf{y}^{(k)} \vee U\mathbf{y}^{(k-1)} \vee \mathbf{b}, \tag{3.99}$$

and since L is strictly lower triangular, we can write (3.99) equivalently as

$$\mathbf{y}^{(k)} = L^* U \mathbf{y}^{(k-1)} \vee L^* \mathbf{b}. \tag{3.100}$$

(Indeed, since L is strictly lower triangular, this matrix is nilpotent, and therefore the system (3.99) has a unique solution—see Section 3.6.1.)

This method is the counterpart of the Gauss–Seidel method of solving a system $\mathbf{y} = A\mathbf{y} + \mathbf{b}$ in linear algebra (Varga 1962). The concrete interpretation of this method for the algebra P_2 is also well known, as *Ford's method* of finding shortest paths (Ford and Fulkerson 1962).

(iii) *The 'double-sweep' method.* The performance of the Gauss–Seidel method may be strongly affected by the ordering of the equations in the system $\mathbf{y} = A\mathbf{y} \vee \mathbf{b}$. For instance, it is evident from (3.97) that if A is strictly lower triangular, the Gauss–Seidel method becomes the forward substitution method of Section 3.6.1, and only one iteration is required. However, if the ordering of the y-variables of the system is reversed, making the A-matrix upper triangular, the Gauss–Seidel method (3.100) becomes equivalent to the Jacobi method (3.95) and $\mathbf{y}^{(k)}$ is constructed using only elements of the *previous* estimate $\mathbf{y}^{(k-1)}$; in this case, a large number of iterations may be needed to obtain the solution.

This defect is largely overcome in the 'double-sweep' iterative method, where initially we choose some estimate $\mathbf{y}^{(0)} \geqslant \mathbf{b}$ and then construct successively the vectors $\mathbf{y}^{(k-\frac{1}{2})}$ and $\mathbf{y}^{(k)}$, $(k = 1, 2, \ldots)$ which are the (unique) solutions of the equations

$$\left.\begin{aligned} \mathbf{y}^{(k-\frac{1}{2})} &= U\mathbf{y}^{(k-\frac{1}{2})} \vee \mathbf{y}^{(k-1)}, \\ \mathbf{y}^{(k)} &= L\mathbf{y}^{(k)} \vee \mathbf{y}^{(k-\frac{1}{2})}, \end{aligned}\right\} \qquad \begin{aligned} &\text{(3.101(a))} \\ &\text{(3.101(b))} \end{aligned}$$

these being respectively

$$\left.\begin{aligned} \mathbf{y}^{(k-\frac{1}{2})} &= U^*\mathbf{y}^{(k-1)}, \\ \mathbf{y}^{(k)} &= L^*\mathbf{y}^{(k-\frac{1}{2})}. \end{aligned}\right\} \qquad \begin{aligned} &\text{(3.102(a))} \\ &\text{(3.102(b))} \end{aligned}$$

The equation (3.101(a)) is solved by the *back*-substitution method of Section 3.6.1, and (3.101(b)) is solved by the *forward*-substitution method. Thus, the elements of $\mathbf{y}^{(k-\frac{1}{2})}$ are calculated in the order $y_n^{(k-\frac{1}{2})}, y_{n-1}^{(k-\frac{1}{2})}, \ldots, y_1^{(k-\frac{1}{2})}$ using

$$y_n^{(k-\frac{1}{2})} = y_n^{(k-1)}, \qquad y_i^{(k-\frac{1}{2})} = \left(\bigvee_{j=i+1}^{n} a_{ij} y_j^{(k-\frac{1}{2})} \right) \vee y_i^{(k-1)},$$

$$(i = n-1, n-2, \ldots, 1), \qquad \text{(3.103(a))}$$

and then the elements of $\mathbf{y}^{(k)}$ are calculated in the order $y_1^{(k)}, y_2^{(k)}, \ldots, y_n^{(k)}$ using

$$y_1^{(k)} = y_1^{(k-\frac{1}{2})}, \qquad y_i^{(k)} = \left(\bigvee_{j=1}^{i-1} a_{ij} y_j^{(k)} \right) \vee y_i^{(k-\frac{1}{2})},$$

$$(i = 2, 3, \ldots, n). \qquad (3.103(\text{b}))$$

It is evident from (3.101) that

$$\mathbf{y}^{(0)} \leqslant \mathbf{y}^{(\frac{1}{2})} \leqslant \mathbf{y}^{(1)} \leqslant \mathbf{y}^{(1+\frac{1}{2})} \leqslant \cdots. \qquad (3.104)$$

Also, from (3.102), we may express this iterative method as

$$\mathbf{y}^{(k)} = L^* U^* \mathbf{y}^{(k-1)} \qquad (3.105)$$

or, since $\mathbf{b} \leqslant \mathbf{y}^{(0)}$, and (3.104) holds, as

$$\mathbf{y}^{(k)} = L^* U^* \mathbf{y}^{(k-1)} \vee L^* U^* \mathbf{b}. \qquad (3.106)$$

This algorithm is a generalization of a method developed by Yen (1970, 1975) for solving shortest-path problems. It does not have a counterpart in linear algebra, although it bears some resemblance to Aitken's 'double-sweep' method (Varga 1962).

3.7.2. Conditions for validity of the iterative methods

Let us suppose that we can represent the closure of the A-matrix of the system (3.92) by a product

$$A^* = M^* N \qquad (3.107)$$

where M and N are $n \times n$ matrices, the matrix M being stable. Then we associate with this product an iterative scheme

$$\mathbf{y}^{(k)} = M \mathbf{y}^{(k-1)} \vee N \mathbf{b}, \qquad (k = 1, 2, \ldots). \qquad (3.108)$$

It will be observed that if (3.107) holds, the equations (3.92) and (3.108) have the same least solutions.

In fact, all the iterative methods of the previous section can be described in this way, their M and N matrices being as follows (cf. (3.95), (3.100), and (3.106)):

Jacobi method: $\qquad M_J = L \vee U \qquad N_J = E \qquad (3.109(\text{a}))$

Gauss–Seidel method: $M_{GS} = L^* U \qquad N_{GS} = L^* \qquad (3.109(\text{b}))$

Double-sweep method: $M_{ds} = L^* U^* \qquad N_{ds} = L^* U^* \qquad (3.109(\text{c}))$

For the Jacobi method the condition (3.107) obviously holds. For the Gauss–Seidel method it follows by (3.27) that when L^*U is stable,

$$M^*_{GS}N_{GS} = (L^*U)^*L^* = (L \vee U)^* = A^*,$$

while for the double-sweep method it follows by (3.20) and (3.26) that when L^*U^* is stable,

$$M^*_{ds}N_{ds} = (L^*U^*)^*L^*U^* = E \vee (L^*U^*)^*L^*U^* = (L^*U^*)^*$$
$$= (L \vee U)^* = A^*.$$

Now for any iterative method of the form (3.108), it follows by substitution in (3.108) that

$$\mathbf{y}^{(k)} = M(M\mathbf{y}^{(k-2)} \vee N\mathbf{b}) \vee N\mathbf{b} = M^2\mathbf{y}^{(k-2)} \vee (E \vee M)N\mathbf{b}$$

and by repeated substitutions,

$$\mathbf{y}^{(k)} = M^k\mathbf{y}^{(0)} \vee (E \vee M \vee M^2 \vee \cdots \vee M^{k-1})N\mathbf{b},$$
$$(k = 1, 2, \ldots). \quad (3.110)$$

Hence, by (3.107),

$$\mathbf{y}^{(k)} = M^k\mathbf{y}^{(0)} \vee A^*\mathbf{b} \quad \text{for all } k \geq q+1, \quad (3.111)$$

where q is the stability index of M, and therefore

$$\mathbf{y}^{(k)} \geq A^*\mathbf{b} \quad \text{for all } k \geq q+1. \quad (3.112)$$

From (3.111) it is clear that in general, the validity of iterative methods of the form (3.108) is not assured for all initial estimates $\mathbf{y}^{(0)}$. However, let us impose the condition

$$\mathbf{y}^{(0)} \leq A^*\mathbf{b}.$$

(In practice this condition is easily met, for instance by setting $\mathbf{y}^{(0)} = \mathbf{b}$.) Then in (3.110) we have

$$M^k\mathbf{y}^{(0)} \leq M^kA^*\mathbf{b} = M^kM^*N\mathbf{b} \leq M^*N\mathbf{b} = A^*\mathbf{b}$$

and therefore

$$\mathbf{y}^{(k)} \leq A^*\mathbf{b}, \quad (k = 1, 2, \ldots). \quad (3.113)$$

Combining (3.112) and (3.113) we obtain

$$\mathbf{y}^{(k)} = A^*\mathbf{b} \quad \text{for all } k \geq q+1.$$

Hence, *any iterative method of the form* (3.108), *with* $\mathbf{y}^{(0)} \leqslant A^*\mathbf{b}$, *gives the least solution of a system* $\mathbf{y} = A\mathbf{y} \vee \mathbf{b}$ *after at most* $q + 1$ *iterations, where q is the stability index of M.*

In the particular case where A is absorptive, this result leads to a useful practical bound to the number of iterations. It was implied by our choice of M and N in (3.107) that

$$M \leqslant MA^* \leqslant M^*A^* = A^*.$$

Hence if A is absorptive then M is also absorptive (see Section 3.4.4) and therefore the stability index of M is less than n. Thus *if A is absorptive, then any method of the form* (3.108), *with* $\mathbf{y}^{(0)} \leqslant A^*\mathbf{b}$, *gives the least solution of a system* $\mathbf{y} = A\mathbf{y} \vee \mathbf{b}$ *after at most n iterations* (*where n is the order of A*).

Given further information about M and N, it is sometimes possible to obtain better upper bounds for the number of iterations. In particular, this can be done for the double-sweep method: by comparing terms in their expansions it is clear that

$$(E \vee L \vee U)^{2k-1} \leqslant (L^*U^*)^k \leqslant A^*, \qquad (k = 1, 2, \ldots). \tag{3.114}$$

Now if A is stable of index q, then†

$$(E \vee L \vee U)^{2k-1} = (E \vee A)^{2k-1}$$
$$= E \vee A \vee A^2 \vee \cdots \vee A^{2k-1} = A^*,$$
$$\text{for all } k \geq \lfloor q/2 \rfloor + 1, \tag{3.115}$$

and by combining (3.114) and (3.115) we obtain

$$(L^*U^*)^k = A^* \quad \text{for all } k \geq \lfloor q/2 \rfloor + 1.$$

It follows from (3.105) that for the double-sweep method, with $\mathbf{b} \leqslant \mathbf{y}^{(0)} \leqslant A^*\mathbf{b}$,

$$\mathbf{y}^{(k)} = (L^*U^*)^k\mathbf{y}^{(0)} = A^*\mathbf{b} \quad \text{for all } k \geq \lfloor q/2 \rfloor + 1.$$

Consequently *the double-sweep method, with* $\mathbf{b} \leqslant \mathbf{y}^{(0)} \leqslant A^*\mathbf{b}$, *gives the least solution of a system* $\mathbf{y} = A\mathbf{y} \vee \mathbf{b}$ *after at most* $\lfloor q/2 \rfloor + 1$ *iterations* (*where q is the stability index of A*). It follows that *if A is absorptive, the method requires at most* $\lfloor (n+1)/2 \rfloor$ *iterations.*

---

† The symbolism $\lfloor x \rfloor$ means "the largest integer not exceeding x."

3.7.3. *Comparison of the iterative methods*

By comparing the definitions (3.96), (3.97), and (3.103) of the Jacobi, Gauss–Seidel, and double-sweep methods we see that in each iteration, the numbers of join and multiplicative operations performed are precisely the same in all of them.

It is also possible to relate the numbers of iterations required, by the following argument. Let us suppose that we have two iterative methods of the form (3.108), the first constructing vectors $\mathbf{y}_1^{(k)}$, $(k = 1, 2, \ldots)$ using

$$\mathbf{y}_1^{(k)} = M_1 \mathbf{y}_1^{(k-1)} \vee N_1 \mathbf{b} \qquad (3.116(\mathrm{a}))$$

and the second constructing vectors $\mathbf{y}_2^{(k)}$, $(k = 1, 2, \ldots)$ where

$$\mathbf{y}_2^{(k)} = M_2 \mathbf{y}_2^{(k-1)} \vee N_2 \mathbf{b}. \qquad (3.116(\mathrm{b}))$$

We assume that the same initial estimate is used in both cases,

$$\mathbf{y}_1^{(0)} = \mathbf{y}_2^{(0)} \qquad (3.117)$$

and that this vector is not greater than $A^*\mathbf{b}$.

Now let us suppose that

$$M_1 \leqslant M_2 \quad \text{and} \quad N_1 \leqslant N_2. \qquad (3.118)$$

Then, since the join and multiplicative operations are isotone, it follows from (3.117) and (3.118) that $\mathbf{y}_1^{(1)} \leqslant \mathbf{y}_2^{(1)}$, which in turn implies that $\mathbf{y}_1^{(2)} \leqslant \mathbf{y}_2^{(2)}$, and so on. Combining this result with (3.113), we obtain

$$\mathbf{y}_1^{(k)} \leqslant \mathbf{y}_2^{(k)} \leqslant A^*\mathbf{b}, \qquad (k = 0, 1, 2, \ldots).$$

Hence, if for some value of k we have $\mathbf{y}_1^{(k)} = A^*\mathbf{b}$, then $\mathbf{y}_2^{(k)} = A^*\mathbf{b}$ also, which implies that the number of iterations required by the second method is not greater than the number required by the first.

For the iterative methods presented in Section 3.7.1 we have (cf. (3.109))

$$M_\mathrm{J}, M_\mathrm{GS} \leqslant M_\mathrm{ds} \quad \text{and} \quad N_\mathrm{J}, N_\mathrm{GS} \leqslant N_\mathrm{ds}.$$

Hence, *for any initial estimate* $\mathbf{y}^{(0)}$ *in the range* $\mathbf{b} \leqslant \mathbf{y}^{(0)} \leqslant A^*\mathbf{b}$, *the number of iterations required by the double-sweep method is not greater than the numbers required by the Jacobi and Gauss–Seidel methods.*

Example 3.26. *Shortest path calculations.* For the shortest-path problem of Example 3.23, the successive **y**-vectors obtained by Yen's double-sweep method, with $\mathbf{y}^{(0)} = \mathbf{b}$, are given below. It will be seen that only two complete iterations are needed to reach the solution. (With any problem, the iterative process can be terminated as soon as two successive **y**-vectors are identical: indeed, by premultiplying (3.102(b)) by L^*, and noting that $L^*L^* = L^*$, we find that $L^*\mathbf{y}^{(k)} = \mathbf{y}^{(k)}$, for $k = 1, 2, \ldots$; hence if $\mathbf{y}^{(k-\frac{1}{2})} = \mathbf{y}^{(k-1)}$, it follows by (3.102(b)) that $\mathbf{y}^{(k)} = L^*\mathbf{y}^{(k-1)} = \mathbf{y}^{(k-1)}$. Similarly, it follows from (3.102(a)) that if $\mathbf{y}^{(k)} = \mathbf{y}^{(k-\frac{1}{2})}$ then all subsequent **y**-vectors have this value.)

$\mathbf{y}^{(0)}$	$\mathbf{s}^{(0)}$	$\mathbf{y}^{(\frac{1}{2})}$	$\mathbf{s}^{(\frac{1}{2})}$	$\mathbf{y}^{(1)}$	$\mathbf{s}^{(1)}$	$\mathbf{y}^{(1+\frac{1}{2})}$	$\mathbf{s}^{(1+\frac{1}{2})}$	$\mathbf{y}^{(2)}$	$\mathbf{s}^{(2)}$
∞	0	∞	0	∞	0	⑥	③	6	3
∞	0	⑧	④	8	4	8	4	⑦	①
∞	0	∞	0	⑨	②	③	⑤	3	5
0	0	0	0	0	0	0	0	0	0
∞	0	∞	0	②	④	2	4	2	4

To obtain the sequences of nodes traversed by shortest paths, we can construct a sequence of *successor vectors* $\mathbf{s}^{(0)}$, $\mathbf{s}^{(\frac{1}{2})}$, $\mathbf{s}^{(1)}$, ... in the course of the computation as follows. In the initial successor vector $\mathbf{s}^{(0)}$, all entries are zero. In the first part of the kth iteration, when each element $y_i^{(k-\frac{1}{2})}$ is formed we set

$$s_i^{(k-\frac{1}{2})} = \begin{cases} s_i^{(k-1)} & \text{if } y_i^{(k-\frac{1}{2})} = y_i^{(k-1)} \\ j & \text{if } y_i^{(k-\frac{1}{2})} \neq y_i^{(k-1)} \end{cases} \qquad (i = 1, 2, \ldots, n),$$

where j is any index such that $y_i^{(k-\frac{1}{2})} = a_{ij}y_j^{(k-1)}$. In the following half iteration the elements of $\mathbf{s}^{(k)}$ are obtained from $\mathbf{s}^{(k-\frac{1}{2})}$ in the same manner:

$$s_i^{(k)} = \begin{cases} s_i^{(k-\frac{1}{2})} & \text{if } y_i^{(k)} = y_i^{(k-\frac{1}{2})} \\ j & \text{if } y_i^{(k)} \neq y_i^{(k-\frac{1}{2})} \end{cases} \qquad (i = 1, 2, \ldots, n),$$

where j is any index such that $y_i^{(k)} = a_{ij}y_j^{(k-\frac{1}{2})}$. It is evident that on termination, the ith element of the final successor vector is the index j of the successor x_j of node x_i on a shortest path from x_i to the destination node. In this way, the final successor vector defines shortest paths from all nodes to the destination.

As an illustration, to find a shortest path from x_1 to x_4 in our example, we find in turn the entries $s_1^{(2)} = 3$, $s_3^{(2)} = 5$, $s_5^{(2)} = 4$, which give the path $x_1 \rightarrow x_3 \rightarrow x_5 \rightarrow x_4$.

3.8. A special method for totally ordered path algebras

Here we consider the particular case of a path algebra P in which (i) the ordering $\leqslant$ is total, and (ii) the unit element e is the greatest element of P.

As examples we have the two-element Boolean algebra, and the algebras P_4 and P_5 of Table 3.1. The conditions (i) and (ii) are also satisfied by the algebra P_2^+ which is obtained from P_2 by replacing the set $\mathbb{R}$ of real numbers by the set of non-negative real numbers. (This can be used to formulate shortest-path problems, when all arc lengths are non-negative.)

Dijkstra's method

Let us again suppose that we are given a system

$$\mathbf{y} = A\mathbf{y} \vee \mathbf{b} \tag{3.119}$$

for which we require the least solution, and let us write this solution as

$$\tilde{\mathbf{y}} = A^*\mathbf{b}. \tag{3.120}$$

Since P is totally ordered, the vector $\mathbf{b}$ has a greatest element, b_k say:

$$b_k \geqslant b_j, \qquad (j = 1, 2, \ldots, n). \tag{3.121}$$

From (3.120), the corresponding element $\tilde{y}_k$ of $\tilde{\mathbf{y}}$ can be expressed as

$$\tilde{y}_k = \bigvee_{j=1}^{n} a_{k_j}^* b_j. \tag{3.122}$$

Now since e is the greatest element of P,

$$a_{kj}^* \leqslant e, \qquad (j = 1, 2, \ldots, n), \tag{3.123}$$

and furthermore,

$$a_{kk}^* = e. \tag{3.124}$$

From (3.121), (3.123) and (3.124),

$$a_{kk}^* b_k \geqslant a_{kj}^* b_j, \qquad (j = 1, 2, \ldots, n),$$

and therefore, from (3.122),

$$\tilde{y}_k = a_{kk}^* b_k = b_k.$$

Having found $\tilde{y}_k$, we can delete the kth equation of the original system (3.199), and substitute b_k for y_k in the remaining ones. In this way the original system, which can be written in expanded form as

$$y_i = \left(\bigvee_{j=1}^{n} a_{ij}y_j \right) \vee b_i, \qquad (i = 1, 2, \ldots, n), \qquad (3.125)$$

is transformed into the system

$$y_i = \left(\bigvee_{\substack{j=1 \\ j \neq k}}^{n} a_{ij}y_j \right) \vee (a_{ik}b_k \vee b_i), \qquad (i = 1, 2, \ldots, k-1, k+1, \ldots, n),$$
$$(3.126)$$

which is of order $n - 1$. Repeating the process, we can find another element of $\tilde{\mathbf{y}}$, and then transform the system (3.126) into a system of order $n - 2$; continuing in this fashion, we eventually obtain all the elements of $\tilde{\mathbf{y}}$.

The method can be implemented by the following algorithm, in which the set of variables $\tilde{y}_i$ which remain undetermined at each stage is defined by the set M of their indices:

Step 1   Let $M = \{1, 2, \ldots, n\}$.

Step 2   Find a greatest element b_k of the set $\{b_i \mid i \in M\}$, and remove the index k from M.

Step 3   If M is empty then go to *End*.

Step 4   For each index $i \in M$, replace b_i by $a_{ik}b_k \vee b_i$. Return to *Step 2*.

End

On termination, the vector $\mathbf{b}$ has been transformed into the required solution $\tilde{\mathbf{y}}$.

This algorithm is a generalized form of Dijkstra's algorithm for finding shortest paths (Dijkstra 1959).

In practice, the work involved in finding a greatest element b_k in each $\mathbf{b}$-vector may be considerable. However, in some problems the *value* of b_k is known *a priori*, in which case the algorithm can be greatly simplified: as an illustration, the simple technique described in Section 2.4 for finding accessible sets is essentially an implementation of Dijkstra's algorithm, on the two-element Boolean algebra. The following example also demonstrates this point.

Example 3.27. *Finding lowest-order paths.* The problem of finding paths of lowest order from a node x_r to all other nodes of a graph can be treated as a shortest path problem, by assigning lengths of '1' to all arcs. Then, using the path algebra P_2^+ defined above, the orders of lowest-order paths to all nodes are given by the least solution of the system $\mathbf{y} = A'\mathbf{y} \vee \mathbf{e}_r$.

The application of Dijkstra's method to this problem has a simple graphical interpretation. If we regard the assignment of a value to $\tilde{y}_k$ as a 'labelling' of the node x_k with the value of $\tilde{y}_k$, then we proceed as follows: first, we label node x_r with a '0'. Then, we attach the label '1' to every successor of the node labelled '0'. Next, we attach the label '2' to all unlabelled successors of nodes labelled '1', and so on until all nodes have been labelled.

When the labelling process is finished, a lowest-order path from x_r to any other node x_s can be found, by tracing a path 'backwards' from x_s in such a way that, at each successive node encountered, the value of the node label decreases. This algorithm has been used extensively in the design of printed circuits (Lee 1961).

3.9. Practical considerations

3.9.1. Implementation of the path-finding algorithms

In most practical problems the number of arcs in a graph is much smaller than the maximum possible number, and consequently its adjacency matrix is *sparse*, that is to say, only a small proportion of the elements of the matrix are non-null. Under these circumstances, the performance of any path-finding program depends strongly on the extent to which it exploits sparsity.

To implement the substitution methods of Section 3.6.1, the iterative methods of Section 3.7 and Dijkstra's method on a computer, it is convenient to specify each row of an A-matrix by a list of its non-null entries, together with their column indices. In this way, one easily avoids storing or operating with any null elements.

In the Gauss and Jordan elimination methods, it is rather more difficult to exploit sparsity because the successive eliminations tend to *fill in* the A-matrix: it is clear from (3.77) and (3.86) that at each elimination step, elements which are null in $A^{(k-1)}$ may become non-null in $A^{(k)}$. However, the techniques which have been devised to overcome this problem in the context of numerical linear algebra are all applicable here. In particular, by using 'linked-list' representations of matrices (Knuth 1968) one can avoid storing and operating on any null elements. The 'optimally ordered elimination schemes' of linear algebra—which essentially permute the rows and columns of a matrix in such a way as to minimize the filling-in of its null entries—are also directly applicable to our problems (Rose 1972; Tewarson 1973).

With regard to the implementation of Dijkstra's method, it is sometimes profitable to use sorting techniques, in the search for greatest elements of the **b**-vectors (Johnson 1972).

3.9.2. *Comparison of the algorithms*

Table 3.3 gives the best available upper bounds of the numbers of join and multiplicative operations required by various methods, to solve a system $\mathbf{y} = A\mathbf{y} \vee \mathbf{b}$ where A is absorptive. To obtain these bounds it has been assumed that: (i) no join or multiplicative operations are performed with null elements in the substitution methods, iterative methods, or Dijkstra method; (ii) in the direct methods, maximum possible fill-in occurs (so that in effect, all entries of the original $A^{(0)}$ matrix are non-null); and (iii) the iterative methods take the maximum possible number of iterations.

TABLE 3.3

Method	Number of ∨-operations	Number of multiplications
Substitution methods	mq	mq
Gauss elimination	$\leq n(n-1)(2n-1)/6$ $+ mn(n-1)$	$\leq n(n-1)(2n-1)/6$ $+ mn(n-1)$
Jordan elimination	$\leq n(n-1)^2/2 + mn(n-1)$	$\leq n(n-1)^2/2 + mn(n-1)$
Matrix closure algorithm	$\leq n(n-1)^2$	$\leq n(n-1)^2$
Jacobi and Gauss–Seidel	$\leq mqn$	$\leq mqn$
Double-sweep (Yen)	$\leq mq\lfloor(n+1)/2\rfloor$	$\leq mq\lfloor(n+1)/2\rfloor$
Dijkstra	$\leq m(n(n-1)/2+q)$	$\leq mq$

n represents the order of A; m is the number of different **b**-vectors for which solutions are required; and q is the number of non-null entries in A.

It must be emphasized however that for sparse matrices, the assumption (ii) above is not realistic; indeed as is pointed out by Rose (1972), when a matrix is sparse its order n has only minor importance, as a measure of the work required by a direct method. Also, in connection with assumption (iii), the number of iterations required by an iterative method is usually small in comparison with the maximum possible number. Thus, the bounds in Table 3.3 must not be regarded as indicators of the relative efficiencies of the various algorithms, when applied to real problems.

Nevertheless, from analytic considerations, and practical experiments (Fontan 1974), we can draw the following general conclusions:

(i) When A is sparse, and solutions are required for only a few **b**-vectors, the double-sweep method is usually the most efficient. It is easy to programme, and its storage requirements are minimal.

(ii) If A is sparse and the disposition of its non-null entries is such that the fill-in can be small (as is the case for instance if A is a 'band' matrix—see Tewarson (1973)), then the Gauss elimination method may involve least work. It becomes more advantageous as the number of **b**-vectors increases, since the transformation of A into an upper triangular matrix need not be repeated. Gaussian elimination involves less work than the Jordan method (Fox 1964; Tewarson 1973). The programming involved, to exploit sparsity, is relatively difficult.

(iii) Dijkstra's method becomes more competitive as the number q of non-null elements in A increases.

With regard to the time complexities of these algorithms, it will be noted that for the path algebras P_1-P_5 the times required to perform a join operation or a multiplication are bounded by constants, so the algorithms are all 'fast'. However, for the algebras P_6-P_8, the time needed to perform a multiplication can increase exponentially with problem size.

Exercises

3.1. Let A be a stable matrix on a path algebra, and let the matrices A and A^* be partitioned in the same manner, into

$$A = \begin{bmatrix} A_{11} & A_{12} \\ A_{21} & A_{22} \end{bmatrix} \quad \text{and} \quad A^* = B = \begin{bmatrix} B_{11} & B_{12} \\ B_{21} & B_{22} \end{bmatrix},$$

where the diagonal submatrices are square. Prove that (on condition that all starred terms are stable) the submatrices of A^* are related to the submatrices of A by the following identities:

$$B_{22} = (A_{21}(A_{11})^*A_{12} \vee A_{22})^*,$$

$$B_{12} = (A_{11})^*A_{12}B_{22},$$

$$B_{21} = B_{22}A_{21}(A_{11})^*,$$

$$B_{11} = (A_{11})^*(A_{12}B_{21} \vee E_{11}),$$

where E_{11} is a unit matrix.

(Note: These identities taken in order define a procedure for calculating A^* which is analogous to the method of inverting a matrix by partitioning.)

3.2 Let P be a path algebra whose multiplicative operation has the cancellation property. Prove that every stable matrix of $M_n(P)$ is absorptive.

3.3 A project involves ten activities, for which the durations and constraints (as described in Example 2.11) are specified in the table below. Find the critical paths, and the slack time for each activity, if the project is to be completed in 60 time units.

Activity	Duration	Predecessors
a	25	—
b	14	a, d, e
c	5	—
d	4	c
e	16	d, h
f	3	a, e
g	10	c
h	20	g
i	2	f
j	6	g

3.4. A reactor in a chemical plant is shut down annually and either overhauled or replaced. The cost of overhaul is related to the age of the reactor, as shown in the table below

Reactor age (years)	1	2	3
Cost of overhaul (£1000s)	10	25	60

The cost of a new reactor is £100 000. The expected life of the whole plant is ten years, starting from new. Determine an optimal policy for the replacement of the reactor over the ten-year period, and the total expenditure under this policy.

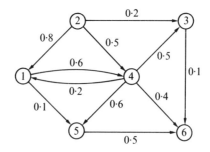

FIG. 3.7

3.5. Solve the following path problems for the graph of Fig. 3.7, using the double-sweep iterative method.

(i) Assuming that the arc labels represent physical length, find shortest paths from node 2 to each of the other nodes.
(ii) Assuming that the label on each arc represents its capacity, find paths of maximum capacity from node 2 to each of the other nodes.
(iii) Assuming that the label on each arc represents its reliability, find a most reliable path from node 2 to each of the other nodes.

3.6. Find all the Hamiltonian cycles on the graph of Fig. 3.8,

(i) by the Gauss elimination method and
(ii) by the double-sweep iterative method.

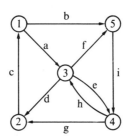

FIG. 3.8

3.7. (i) Prove that, with an initial approximation $\mathbf{y}^{(0)} = \mathbf{b}$, the $\mathbf{y}^{(k)}$-vectors obtained by the Jacobi method (3.109(a)), the Gauss–Seidel method (3.109(b)), and the double-sweep iterative method (3.109(c)) are respectively

$$\mathbf{y}_J^{(k)} = (E \vee L \vee U)^k \mathbf{b}, \qquad \mathbf{y}_{GS}^{(k)} = (L^* \vee L^* U)^k \mathbf{b}, \qquad \mathbf{y}_{ds}^{(k)} = (L^* U^*)^k \mathbf{b}.$$

(ii) Using the results of (i) above, show that with $\mathbf{y}^{(0)} = \mathbf{b}$, the number of iterations needed to solve an equation $\mathbf{y} = A\mathbf{y} \vee \mathbf{b}$ by the Gauss–Seidel method is not greater than the number required by the Jacobi method.

(iii) Compare the $\mathbf{y}^{(k)}$-vectors obtained by the three methods in the particular cases where (a) the matrix A is strictly lower triangular, (b) A is strictly upper triangular, (c) A is of the form

$$A = \begin{bmatrix} \Phi_1 & B \\ C & \Phi_2 \end{bmatrix}$$

where the null submatrices Φ_1 and Φ_2 are both square.

3.8. (i) Prove that for any elements w, x, y, z of a path algebra

$$(w \vee xyz)^* = w^* \vee w^*x(yzw^*x)^*yzw^*.$$

(ii) Let A be a square matrix on a path algebra, and let B be the matrix obtained by changing the value of some element a_{ij} of A to σ, where $\sigma \geqslant a_{ij}$. Using the identity given above, prove that

$$B^* = A^* \vee \mathbf{c}(\sigma a_{ji}^*)^* \sigma \mathbf{d},$$

where $\mathbf{c}$ is the ith column of A^* and $\mathbf{d}$ is the jth row of A^*.

3.9. Let P be a path algebra whose multiplicative operation is commutative. Then the *determinant* $|A|$ of a matrix $A \in M_n(P)$ is defined by

$$|A| = \bigvee_{h_1,\ldots,h_n} a_{1h_1}a_{2h_2} \ldots a_{nh_n}$$

where $\bigvee_{h_1,\ldots,h_n}$ means that the join is extended over all permutations $(h_1, h_2, \ldots, h_n)$ of the indices $1, 2, \ldots, n$. The *adjoint* of A is the matrix $adj\ (A) = [\alpha_{ij}]$ with elements

$$\alpha_{ij} = |A_{ji}| \quad \text{for all } i, j,$$

where $|A_{ji}|$ is the determinant of the minor A_{ji} of a_{ji} in A.
Prove that if A is absorptive then

$$adj\ (E \vee A) = A^*.$$

3.10. (i) Let G be an n-node graph labelled with the path algebra P_2 (for finding shortest paths), let A be the adjacency matrix of G, and let $A^{(k)} = [a_{ij}^{(k)}]$ be the matrix obtained at the kth step of the Gauss elimination method (as defined by (3.81)). Prove that if the subgraph of G generated by $\{x_1, x_2, \ldots, x_k\}$ does not contain a cycle of negative length, then each non-null entry $a_{ij}^{(k)}$ of $A^{(k)}$ for which $i > k$ and $j > k$ is the length of a shortest elementary path from x_i to x_j, on the subgraph of G generated by $\{x_1, x_2, \ldots, x_k\} \cup \{x_i, x_j\}$.

(ii) Using this result, develop an algorithm to determine whether a given graph contains any cycles of negative length. Devise also a method for finding a cycle of negative length, if one exists.

Additional notes and bibliography

The first algebraic study of path problems was by Lunts (1950), who applied Boolean matrix algebra to the analysis of relay networks. Moisil (1960) and Yoeli (1961) extended his results to a more general algebraic structure, applicable to several different types of path problems, and further extensions of this work were described subsequently by Cruon and Hervé (1965), Tomescu (1966, 1968), Peteanu (1967, 1969, 1970), Benzaken (1968), Robert and Ferland (1968), Carré (1971), Derniame and Pair (1971), Backhouse and Carré (1975), Gondran (1975), Roy (1975), and Wongseelashote (1976). See also Cuninghame-Green (1962, 1976) and Minoux (1976).

The 'path algebra' presented in Section 3.2 will be recognized by algebraists as a multiplicative semi-lattice, or semi-lattice-ordered monoid. For a fundamental treatment of lattice-ordered structures see Birkhoff (1967) and Dubreil-Jacotin, Lesieur, and Croisot (1953).

Some further concrete examples of the path algebra will be found in Derniame and Pair (1971), Minieka and Shier (1973), Brucker (1974), Gondran (1975) and Shier (1976).

The formulae (3.71), (3.72), and (3.83) defining the Gauss and Jordan elimination methods suggest the possibility of developing triangular factors and product forms of closure matrices, analogous to the triangular factorizations and product forms of inverses used in ordinary matrix algebra; these are described by Carré (1969, 1971) and Backhouse and Carré (1975). Many other techniques of numerical linear algebra are applicable to path problems, and several of them have been reinvented in a graph-theoretic context. For instance, the escalator method of matrix inversion (Faddeeva 1959) has been developed as a method of computing shortest paths by Dantzig (1966). Graph-theoretic counterparts of the 'decomposition' or 'block' methods of inverting sparse matrices (Tewarson 1973) have also been discovered (Land and Stairs 1967; Hu and Torres 1969; Yen 1975); see also Hoffman and Winograd (1972). We note however that in numerical linear algebra, decomposition methods have been largely superceded by ordered elimination techniques (Tewarson 1973; Duff 1977).

The formula given in Exercise 3.8 is our counterpart of the method of finding inverses of modified matrices in linear algebra (Householder 1953). This formula was derived on P_2 from graph-theoretic considerations by Murchland (1967) and Rodionov (1968), who used it to calculate the changes in distances on a graph when one of its arc lengths is reduced.

The solution of path problems using determinants (see Exercise 3.9) was first proposed by Lunts (1950), and Hammer and Rudeanu (1968) used determinants on a Boolean algebra to enumerate elementary paths. There is a close connection between the method of evaluating such determinants, using expansion by elements of rows, and the back-track programming method of finding elementary paths which was presented in Section 2.7.

A bibliography on algorithms for path problems has been published by Pierce (1975).

4 Connectivity

4.1. Introduction

IN THIS CHAPTER we shall first consider the ways in which it is possible to 'separate' two nodes of a graph, that is, to destroy all the paths between them, by removing arcs. The concept of a 'separating arc set' which is developed here is important for instance in determining the reliability of a communication network with faulty links, and in studying flows in transportation networks. Next we shall discuss the complementary notion of a minimal 'connecting' set of arcs of a graph, which preserves the accessibility between nodes, even when all the other arcs are removed. We shall then consider the separation of two points, by removing nodes rather than arcs; this type of decomposition of a graph finds applications for instance in the compilation of computer programs.

After exploring these concepts, in which arc orientations are significant, we shall present their 'undirected' counterparts for simple graphs, whose edges are not oriented.

4.2. Separation by the removal of arcs

4.2.1. Separating arc sets

Let x_i and x_j be any two nodes of a graph $G = (X, U)$. Then a subset V of U is an (x_i, x_j)-*separating arc set* if every path from x_i to x_j traverses at least one arc in V. When V is an (x_i, x_j)-separating arc set, and no proper subsets of V have this property, we say that V is a *proper* (x_i, x_j)-*separating arc set*.

Example 4.1. In the graph of Fig. 4.1, the proper (x_5, x_4)-separating arc sets are

$$\{a, b, h\}, \{b, c, h\}, \{b, d, h\}, \{g, h\}.$$

4.2.2. Cut sets of arcs

Now let x_i and x_j be two distinct nodes on a graph $G = (X, U)$ and let $\{X', X''\}$ be any partition of the node set of G such that $x_i \in X'$ and $x_j \in X''$. Then the set of all arcs with initial endpoints in X' and

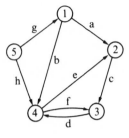

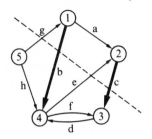

FIG. 4.1   FIG. 4.2

terminal endpoints in X'' is called a *cut set of arcs* (or simply a *cut*) *separating* x_i *from* x_j (in that order). It will sometimes be convenient to denote this cut by (X', X''); thus

$$(X', X'') = \{(x_i, x_j) \in U \mid x_i \in X', x_j \in X''\}.$$

When (X', X'') is a cut separating x_i from x_j, and no proper subsets of (X', X'') have this property, we say that (X', X'') is a *proper cut separating* x_i *from* x_j.

Example 4.2. The following are two examples of cuts separating x_1 from x_4, on the graph of Fig 4.1:

(i) By partitioning the node set X into the subsets $X' = \{x_1, x_2\}$ and $X'' = \{x_3, x_4, x_5\}$ we obtain the cut $(X', X'') = \{b, c\}$. This partition of X is indicated by a broken line in Fig. 4.2; the bold lines indicate the members of the cut.

(ii) With $X' = \{x_1, x_2, x_5\}$ and $X'' = \{x_3, x_4\}$ we obtain the cut $(X', X'') = \{b, c, h\}$. The first of these is a proper cut separating x_1 from x_4; the second is not a proper cut separating x_1 from x_4, since it contains the first.

The relationship between the cuts and the separating arc sets of a graph can be summarized as follows:

(1) *Every cut* (X', X'') *separating* x_i *from* x_j *is an* (x_i, x_j)-*separating arc set*. Indeed, since $x_i \in X'$ and $x_j \in X''$, every path from x_i to x_j must traverse some arc of (X', X'').

The converse of (1) is not necessarily true. For instance, on a complete symmetric graph the set of all arcs is a separating arc set, for every pair of nodes, but it is not a cut. However we can prove the following:

(2) *For any two distinct nodes* x_i *and* x_j, *every proper* (x_i, x_j)-*separating arc set is a proper cut separating* x_i *from* x_j. Indeed, let x_i

and x_j be distinct nodes on G, let V be any proper (x_i, x_j)-separating arc set, let X' be the set of nodes which are accessible from x_i on the partial graph of G obtained by removing the arcs of V, and let X'' be the complement of X' relative to X. Then clearly $x_i \in X'$ and $x_j \in X''$, and therefore on G, the arc set (X', X'') is a cut separating x_i from x_j. Now $(X', X'') \subseteq V$, by the definitions of X' and X''; also, since (X', X'') is an (x_i, x_j)-separating arc set (by result (1) above), and V is a proper (x_i, x_j)-separating arc set, (X', X'') cannot be a proper subset of V. It follows that $(X', X'') = V$, which implies that V is a cut separating x_i from x_j. Furthermore, since none of the proper subsets of V are (x_i, x_j)-separating arc sets, it follows from result (1) above that none of these subsets are cuts separating x_i from x_j. Hence V is a proper cut separating x_i from x_j, as required.

From the results (1) and (2) we deduce the following:

(3) *Every proper cut separating x_i from x_j is a proper (x_i, x_j)-separating arc set.* Indeed let C be any proper cut separating x_i from x_j. Then by result (1) above, C is an (x_i, x_j)-separating arc set. Now if C is not a proper (x_i, x_j)-separating arc set, it must contain a subset $C' \subset C$ which does have this property. But then, by result (2), C' is a proper cut separating x_i from x_j, which violates our initial assumption. Hence C is a proper (x_i, x_j)-separating arc set, as required.

The concept of a cut will be used in studying network flows, in Chapter 6.

4.2.3. The determination of proper separating arc sets

The proper separating arc sets on a graph can be found by means of the following path algebra, which is due to Martelli (1974, 1976).

Let Σ be a finite alphabet, and let S be any set of subsets of Σ; then we say that a member σ of S is a *minimal* member of S if S does not contain any proper subsets of σ. The *reduction* $r(S)$ of S is the set of all minimal members of S. If $r(S) = S$ then S is a *reduced* set of sets; in particular, the empty set ϕ and the set $\Phi = \{\phi\}$ are both reduced sets of sets.

Let P_Σ be the set of all reduced sets of subsets of Σ, and let us define a join and multiplicative operation on P_Σ by the rules

$$\left. \begin{array}{l} X \vee Y = r(\{\alpha \cup \beta \,|\, \alpha \in X, \beta \in Y\}), \\ X \cdot Y = r(X \cup Y), \end{array} \right\} \quad \text{for all } X, Y \in P_\Sigma. \qquad (4.1)$$

It is easily verified that with these operations P_Σ forms a path algebra, whose unit and zero elements are the sets ϕ and Φ respectively. It will be noted that the unit element ϕ is the greatest element of P_Σ.

Now let G be a graph in which each arc (x_i, x_j) has a name n_{ij}. Let Σ be the set of arc names, and let P_Σ be the path algebra derived from Σ in the manner indicated above. If we assign to each arc (x_i, x_j) the label $\{\{n_{ij}\}\}$, then we may consider G as a graph labelled with P_Σ. In this case, the label $l(\mu)$ of each path μ on G is the union of the labels of its arcs. For instance, on Fig. 4.1 the path $\mu = (x_1, x_4), (x_4, x_3), (x_3, x_4)$ has the label $l(\mu) = \{\{b\}, \{f\}, \{d\}\}$. Since ϕ is the greatest element of P_Σ, all graphs labelled with P_Σ are absorptive.

Now let A be the adjacency matrix of G. Then, by (3.56), its weak closure $\hat{A} = [\hat{a}_{ij}]$ has entries

$$\hat{a}_{ij} = \bigvee_{\mu \in \tilde{T}_{ij}} l(\mu) \tag{4.2}$$

where $\tilde{T}_{ij}$ is the set of all non-null elementary paths from x_i to x_j.

If we denote the elementary paths from x_i to x_j by

$$\tilde{T}_{ij} = \{\mu_{ij}^{(1)}, \mu_{ij}^{(2)}, \ldots, \mu_{ij}^{(q)}\},$$

then (4.2) can be rewritten as

$$\hat{a}_{ij} = \bigvee_{k=1}^{q} l(\mu_{ij}^{(k)}),$$

and therefore from (4.1)

$$\hat{a}_{ij} = r(M_{ij}) \quad \text{where} \quad M_{ij} = \left\{ \bigcup_{k=1}^{q} \alpha_k \mid \alpha_k \in l(\mu_{ij}^{(k)}) \right\}. \tag{4.3}$$

Here M_{ij} is the set of all sets of arc names which can be formed by taking the label of one arc from each elementary path from x_i to x_j. Thus, if x_j is a descendant of x_i then $\hat{a}_{ij}$ is the set of all proper (x_i, x_j)-separating arc sets; whereas if x_j is not a descendant of x_i then $\hat{a}_{ij} = \Phi$.

Example 4.3. The adjacency matrix of the graph of Fig. 4.1. is given in Fig. 4.3(a). To obtain the proper separating arc sets which separate each node from node x_4 say, we require the fourth column of $\hat{A}$. This can be found by solving the system $\mathbf{y} = A\mathbf{y} \vee \mathbf{a}_4$, using any of the direct or iterative methods of Sections 3.6 and 3.7. As an illustration, the successive **y**-vectors obtained by the double-sweep method are given in Fig. 4.3(b). (The solution is obtained in one iteration.)

$$
\begin{bmatrix}
\Phi & \{\{a\}\} & \Phi & \{\{b\}\} & \Phi \\
\Phi & \Phi & \{\{c\}\} & \Phi & \Phi \\
\Phi & \Phi & \Phi & \{\{d\}\} & \Phi \\
\Phi & \{\{e\}\} & \{\{f\}\} & \Phi & \Phi \\
\{\{g\}\} & \Phi & \Phi & \{\{h\}\} & \Phi
\end{bmatrix}
$$

FIG. 4.3(a)

$$
\mathbf{y}^{(0)} \quad \mathbf{y}^{(\frac{1}{2})} \qquad\qquad \mathbf{y}^{(1)}
$$

$$
\begin{bmatrix}
\{\{b\}\} \\
\Phi \\
\{\{d\}\} \\
\Phi \\
\{\{h\}\}
\end{bmatrix}
\begin{bmatrix}
\{\{ab\}, \{bc\}, \{bd\}\} \\
\{\{c\}, \{d\}\} \\
\{\{d\}\} \\
\Phi \\
\{\{h\}\}
\end{bmatrix}
\begin{bmatrix}
\{\{ab\}, \{bc\}, \{bd\}\} \\
\{\{c\}, \{d\}\} \\
\{\{d\}\} \\
\{\{cf\}, \{d\}, \{ef\}\} \\
\{\{abh\}, \{bch\}, \{bdh\}, \{gh\}\}
\end{bmatrix}
$$

FIG. 4.3(b)

An alternative technique for finding proper separating arc sets, using a network flow method, will be presented in Chapter 6.

4.2.4. Basic arcs

An arc $u = (x_i, x_j)$ of a graph G is called a *basic arc* of G if on the graph obtained by removing u from G, x_j is not a descendant of x_i. In other words, u is a basic arc if and only if $\{u\}$ is a proper separating arc set. As an example, the graph of Fig. 4.1 has four basic arcs: c, d, e, and g.

If an arc $u = (x_i, x_j)$ is not a basic arc of G then obviously G contains an elementary path from x_i to x_j which does not traverse u, and which is therefore of order greater than one. For this reason, an arc which is not basic is called a *chord*.

The determination of basic arcs. It is easy to find the basic arcs of a graph, by means of the following path algebra: Let Σ be a finite alphabet, and let ω be any symbol which does not belong to Σ. Let $\Omega = \{\omega\}$, and let

$$
P = \mathscr{P}(\Sigma) \cup \{\Omega\},
$$

where $\mathscr{P}(\Sigma)$ is the power set of Σ. We define a join operation on P by the rules

$$\left.\begin{array}{ll} X \vee Y = X \cap Y & \text{for all } X, Y \in \mathscr{P}(\Sigma), \\ X \vee \Omega = X = \Omega \vee X & \text{for all } X \in P, \end{array}\right\} \tag{4.4}$$

and we define multiplication on P by

$$\left.\begin{array}{ll} X \cdot Y = X \cup Y & \text{for all } X, Y \in \mathscr{P}(\Sigma), \\ X \cdot \Omega = \Omega = \Omega \cdot X & \text{for all } X \in P. \end{array}\right\} \tag{4.5}$$

It is easily verified that the set P equipped with these operations is a path algebra, whose unit and zero elements are the sets ϕ and Ω respectively. It will be noted that the unit element ϕ is the greatest element of P.

Now let G be a graph in which each arc (x_i, x_j) has a name n_{ij}. Let Σ be the set of these names and let P be the path algebra derived from Σ, as indicated above. If each arc (x_i, x_j) is assigned the label $\{n_{ij}\}$, we may consider G as a graph labelled with P. Then for any path μ on G, the label $l(\mu)$ is the union of the labels of the arcs of μ; for instance on Fig. 4.1 the path $\mu = (x_1, x_4), (x_4, x_3), (x_3, x_4)$ has the label $l(\mu) = \{b, f, d\}$. Since ϕ is the greatest element of P, all graphs labelled with P are absorptive.

Let A be the adjacency matrix of G. Then by (3.56) and (4.4), its weak closure $\hat{A}$ has entries

$$\hat{a}_{ij} = \bigvee_{\mu \in \tilde{T}_{ij}} l(\mu) = \begin{cases} \bigcap_{\mu \in \tilde{T}_{ij}} l(\mu) & \text{if } \tilde{T}_{ij} \text{ is not empty,} \\ \Omega & \text{if } \tilde{T}_{ij} \text{ is empty} \end{cases} \tag{4.6}$$

where $\tilde{T}_{ij}$ is the set of non-null elementary paths from x_i to x_j. Thus if x_j is a descendant of x_i, $\hat{a}_{ij}$ is the set of arcs which are common to all elementary paths from x_i to x_j, that is, the set of all basic arcs which separate x_i from x_j; whereas if x_j is not a descendant of x_i then $\hat{a}_{ij} = \Omega$.

The matrix $\hat{A}$ (or any particular row or column of $\hat{A}$) can be computed by any of the direct or iterative methods of Sections 3.6 or 3.7.

Example 4.4. For the graph of Fig. 4.1.

$$A = \begin{bmatrix} \Omega & \{a\} & \Omega & \{b\} & \Omega \\ \Omega & \Omega & \{c\} & \Omega & \Omega \\ \Omega & \Omega & \Omega & \{d\} & \Omega \\ \Omega & \{e\} & \{f\} & \Omega & \Omega \\ \{g\} & \Omega & \Omega & \{h\} & \Omega \end{bmatrix}, \qquad \hat{A} = \begin{bmatrix} \Omega & \phi & \phi & \phi & \Omega \\ \Omega & \{c,d,e\} & \{c\} & \{c,d\} & \Omega \\ \Omega & \{d,e\} & \{d\} & \{d\} & \Omega \\ \Omega & \{e\} & \phi & \{d\} & \Omega \\ \{g\} & \phi & \phi & \phi & \Omega \end{bmatrix}.$$

4.3. Basis graphs

So far we have been interested in removing arcs in such a way as to destroy all the paths between two nodes. Here we again consider the removal of arcs, but our aim is to eliminate arcs which are 'superfluous', in the sense that their removal does *not* change the accessible set of any node.

Given a graph $G = (X, U)$, we describe a partial graph $H = (X, V)$ of G as a *basis graph* of G if

(i) for every node $x_i \in X$, each descendant of x_i on G is a descendant of x_i on H, and
(ii) the arc set of H is minimal (in the sense that if any arc is removed from H, condition (i) is no longer satisfied).

As an example, the graph of Fig. 4.4(a) has a unique basis graph, which is shown in Fig. 4.4(b).

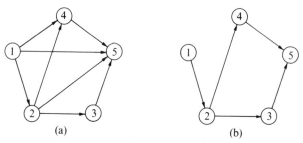

(a)   (b)

FIG. 4.4

It is evident that for any graph, we can always obtain a basis graph, by the successive removal of chords. However, a graph may have several basis graphs, with different numbers of arcs. For instance, the graph of Fig. 4.5(a) has five basis graphs, two of which are shown in Figs. 4.5(b) and 4.5(c).

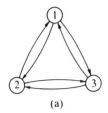
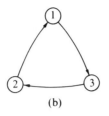
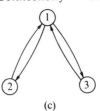

(a)   (b)   (c)

FIG. 4.5

Let us consider the question of when a graph has a unique basis graph. First, we observe that the concept of a basis graph is related to that of a basic arc, as defined in the previous section, in the following manner: *In a graph G, an arc (x_i, x_j) is basic if and only if it belongs to every basis graph of G.* To prove this, let us first suppose that (x_i, x_j) is a basic arc. Then by definition, all the paths from x_i to x_j traverse the arc (x_i, x_j), and therefore this arc must belong to every basis graph of G. Conversely, let us suppose that (x_i, x_j) belongs to every basis graph of G. Then there exist two nodes x_k and x_l such that x_l is a descendant of x_k on G, but not on the partial graph of G obtained by removing (x_i, x_j). Hence on G all the paths from x_k to x_l traverse (x_i, x_j), which implies that all the paths from x_i to x_j traverse (x_i, x_j). Thus (x_i, x_j) is a basic arc, as required.

From this result it follows that a graph G has a unique basis graph if and only if its basic arcs form a basis graph. Hence, *a necessary and sufficient condition for a graph to have a unique basis graph is that for every arc (x_i, x_j) of G, there exists a path from x_i to x_j consisting entirely of basic arcs.*

This condition applies in particular to acyclic graphs. Indeed, let G be an acyclic graph, let (x_i, x_j) be any arc of G, and let μ be any path of maximum order from x_i to x_j. Now let us suppose that some arc (x_k, x_l) of μ is not basic; then G contains a path from x_k to x_l of order greater than one, which contradicts our assumption that μ is of maximum order. Hence μ consists entirely of basic arcs, as required. We conclude therefore that *every acyclic graph has a unique basis graph, consisting of all its basic arcs.*

Example 4.5. *Construction of the basis graph of an acyclic graph.* Since all the paths in an acyclic graph are elementary, an arc (x_i, x_j) of an acyclic graph G is basic if and only if G does not contain *any* paths of order greater than one from x_i to x_j. Hence if G has a Boolean adjacency matrix A, then the Boolean adjacency matrix $\tilde{A}$ of the basis graph of G can be written as[†]

$$\tilde{A} = A \wedge \overline{A\hat{A}}.$$

Here the weak closure $\hat{A}$ of A can be calculated conveniently by Warshall's algorithm (3.91). (To save work, one can exploit the fact that when A is strictly upper triangular, all the $B^{(k)}$ matrices in (3.91) have this property also.)

As an illustration, for the graph of Fig. 4.4(a),

$$A = \begin{bmatrix} 1 & 0 & 1 & 1 \\ & 1 & 1 & 1 \\ & & 0 & 1 \\ \mathbf{0} & & & 1 \end{bmatrix}, \quad \hat{A} = \begin{bmatrix} 1 & 1 & 1 & 1 \\ & 1 & 1 & 1 \\ & & 0 & 1 \\ \mathbf{0} & & & 1 \end{bmatrix},$$

$$A\hat{A} = \begin{bmatrix} 0 & 1 & 1 & 1 \\ & 0 & 0 & 1 \\ & & 0 & 0 \\ \mathbf{0} & & & 0 \end{bmatrix}, \quad \hat{A} = \begin{bmatrix} 1 & 0 & 0 & 0 \\ & 1 & 1 & 0 \\ & & 0 & 1 \\ \mathbf{0} & & & 1 \end{bmatrix}.$$

The corresponding basis graph is shown in Fig. 4.4(b).

An algorithm to perform this calculation has been described by Fisher, Liebman, and Nemhauser (1968), who use it to remove the chords from activity graphs for critical path analysis. (It will be observed that in activity graphs of the kind described in Example 2.11, chords are always 'superfluous' in the sense that they never determine earliest or latest starting times.)

4.4. Separation by the removal of nodes

4.4.1. *Separating node sets*

Let us now consider the ways in which the paths between two nodes can be destroyed, by removing *nodes* rather than arcs.

Let x_i and x_j be any two nodes (which need not be distinct) on a graph $G = (X, U)$. Then an (x_i, x_j)-*separating node set* is a set W of

[†] Given two $m \times n$ Boolean matrices $X = [x_{ij}]$ and $Y = [y_{ij}]$, their *meet* is the $m \times n$ matrix $X \wedge Y = [x_{ij} \wedge y_{ij}]$.

nodes, not containing x_i or x_j, such that every path from x_i to x_j traverses at least one node of W. If W is an (x_i, x_j)-separating node set, and none of its proper subsets have this property, we say that W is a *proper (x_i, x_j)-separating node set*.

It will be noted that if x_j is a successor of x_i, then there do not exist any (x_i, x_j)-separating node sets.

Example 4.6. In the graph of Fig. 4.1, the proper (x_5, x_3)-separating node sets are $\{x_1, x_4\}$ and $\{x_2, x_4\}$.

The determination of proper separating node sets. Separating node sets can be obtained by a technique similar to that used previously to determine separating arc sets. Let us suppose that the nodes of a graph G are numbered $1, 2, \ldots, n$. Let $\Sigma = \{1, 2, \ldots, n\}$, and let P_Σ be the path algebra derived from Σ as in Section 4.2.3. Then, if we give each arc (x_i, x_j) of G the name of its terminal endpoint, and set $l(x_i, x_j) = \{\{j\}\}$, we obtain a graph labelled with P_Σ. It follows by (4.3) that if x_j is accessible from x_i, then the entry $\hat{a}_{ij}$ of the closure matrix $\hat{A}$ is the reduction of the set M_{ij} of all sets of node indices which can be formed by taking the index of one node on each elementary path from x_i to x_j, other than the initial node x_i; thus $\hat{a}_{ij}$ comprises the sets of indices of all the proper (x_i, x_j)-separating node sets, together with the set $\{j\}$. If x_j is not accessible from x_i then $\hat{a}_{ij} = \Phi$.

Example 4.7. For the graph of Fig. 4.1,

$$A = \begin{bmatrix} \Phi & \{\{2\}\} & \Phi & \{\{4\}\} & \Phi \\ \Phi & \Phi & \{\{3\}\} & \Phi & \Phi \\ \Phi & \Phi & \Phi & \{\{4\}\} & \Phi \\ \Phi & \{\{2\}\} & \{\{3\}\} & \Phi & \Phi \\ \{\{1\}\} & \Phi & \Phi & \{\{4\}\} & \Phi \end{bmatrix},$$

$$\hat{A} = \begin{bmatrix} \Phi & \{\{2\}\} & \{\{2, 4\}, \{3\}\} & \{\{4\}\} & \Phi \\ \Phi & \{\{2\}, \{3\}, \{4\}\} & \{\{3\}\} & \{\{3\}, \{4\}\} & \Phi \\ \Phi & \{\{2\}, \{4\}\} & \{\{3\}, \{4\}\} & \{\{4\}\} & \Phi \\ \Phi & \{\{2\}\} & \{\{3\}\} & \{\{3\}, \{4\}\} & \Phi \\ \{\{1\}\} & \{\{1, 4\}, \{2\}\} & \{\{1, 4\}, \{2, 4\}, \{3\}\} & \{\{4\}\} & \Phi \end{bmatrix}.$$

4.4.2. Separating nodes

A node x_k is called an (x_i, x_j)-*separating node* if every path from x_i to x_j traverses x_k. In other words, x_k is an (x_i, x_j)-separating node if and only if $\{x_k\}$ is an (x_i, x_j)-separating node set.

As an illustration, on Fig. 4.6(a), the node x_6 is an (x_1, x_7)-separating node.

Example 4.8. *Dominators of computer programs.* A computer program is often represented by a *control flow graph G* in which each node represents a 'block' of program statements, i.e. a sequence of statements such that (i) no other statement of the program can transfer control to any but the first statement of the sequence, and (ii) if control is passed to the first statement of the sequence then all the statements of the sequence are executed, in order; two nodes x_i and x_j in G are joined by an arc (x_i, x_j) if the last statement of block i can transfer control of the first statement of block j. The node corresponding to the block containing the first statement of the program is called the *entry node* of G, and the node representing the block which contains the 'end' statement of the program is called the *exit node*. We note that if a program is 'properly constructed' then every node of the control flow graph is accessible from its entry node (an inaccessible node would represent a block of statements which could never be executed), and the exit node is accessible from all other nodes (for if control could be passed to a block from which the exit could not be reached, the program would not always terminate).

An important concept in program analysis and code optimization is that of *domination* of one block by another: We say that a block i *pre-dominates* a block j if the first execution of j is always preceded by an execution of i; whereas block i *post-dominates* block j if the last execution of j is always followed by an execution of i. To interpret this notion in graph-theoretic terms, let us denote by x_s and x_t the entry and exit nodes of a control flow graph. Then block i pre-dominates block j if x_i is an (x_s, x_j)-separating node, or $x_i = x_s$; whereas block i post-dominates block j if x_i is an (x_j, x_t)-separating node, or $x_i = x_t$.

For a discussion of the significance of dominators in programming see for instance Schaefer (1973) and Hecht (1977).

An algebraic method of finding separating nodes. Separating nodes can be found by a technique similar to that used previously for finding basic arcs. Let us suppose that the nodes of a graph G are numbered $1, 2, \ldots, n$, and that each arc (x_i, x_j) is assigned the label $l(x_i, x_j) = \{j\}$. Let $\Sigma = \{1, 2, \ldots, n\}$ and let P_Σ be the path algebra derived from Σ as in Section 4.2.4. By (4.6), if a node x_j is accessible from a node x_i, the entry $\hat{a}_{ij}$ of the weak closure matrix $\hat{A}$ is the set of indices of the nodes which belong to every path from x_i to x_j, other than the initial node x_i; thus $\hat{a}_{ij}$ comprises the indices of all the (x_i, x_j)-separating nodes, together with the index 'j'. If x_j is not accessible from x_i then (by (4.6)) $\hat{a}_{ij} = \Omega$.

Example 4.9. Using the technique described above, the adjacency matrix A of the graph of Fig. 4.6(a) has the form shown in Fig. 4.6(b). To find the separating nodes between each node and node x_7 say, we solve the equation $\mathbf{y} = A\mathbf{y} \vee \mathbf{a}_7$ (where $\mathbf{a}_7$ is the seventh column vector of A). Applying the double-sweep method to this problem, we obtain the iterates shown in Fig. 4.6(c).

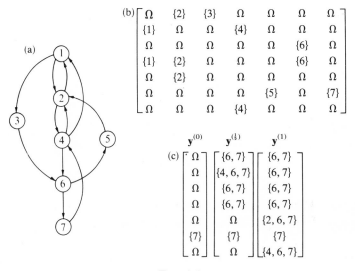

FIG. 4.6

An alternative method of finding separating nodes is outlined in Exercise 4.3. (See also the Additional notes and bibliography.)

4.5. Edge separation on simple graphs

In all the notions of connectivity discussed so far, the orientations of the arcs of a graph have been significant. The remainder of this chapter presents 'undirected' counterparts of these concepts relating to simple graphs (as defined in Section 2.2), in which the edges are not oriented.

4.5.1. *Separating edge sets* (cf. Section 4.2.1)

Let $G = (X, E)$ be a connected simple graph, and let x_i and x_j be any two distinct nodes of G. Then a set F of edges of G is a *separating edge set between x_i and x_j* if every chain between x_i and x_j

contains at least one edge of F. When F is a separating edge set between x_i and x_j, and no proper subset of F has this property, we say that F is a *proper separating edge set* between x_i and x_j.

Example 4.10. In the graph of Fig. 4.7, the proper separating edge sets between x_3 and x_7 are

$$\{a, b, e\}, \{b, d, e\}, \{c, e\}, \{f\}, \{g, h\}, \{h, i\}.$$

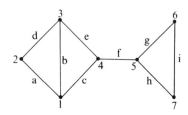

FIG. 4.7

4.5.2. *Cut sets of edges* (cf. Section 4.2.2)

Again, let x_i and x_j be any two distinct nodes of a connected simple graph $G = (X, E)$, and let $\{X', X''\}$ be a partitioning of the node set X such that $x_i \in X'$ and $x_j \in X''$. Then the set F of edges which have one endpoint in X' and the other endpoint in X'' is called a *cut set of edges*, between x_i and x_j.

When F is a cut set of edges, but no proper subset of F has this property, we say that F is a *proper cut set of edges*.

Clearly, a cut set of edges on a connected graph $G = (X, E)$ is proper if and only if the graph $\tilde{G} = (X, E - F)$ has precisely two connected components: For if $\tilde{G}$ has only two components then the addition of any edge of F to $\tilde{G}$ reunites these components, which implies that none of the proper subsets of F are cut sets on G; whereas if $\tilde{G}$ has more than two components, then after adding any edge $f \in F$ to $\tilde{G}$, this graph is still not connected, which implies that the edge set $F - \{f\}$ is a cut set on G.

Example 4.11. As examples of cut sets of edges between x_3 and x_7, on the graph of Fig. 4.7 we have

(i) with $X' = \{x_2, x_3\}$ and $X'' = \{x_1, x_4, x_5, x_6, x_7\}$, the cut $\{a, b, e\}$;
(ii) with $X' = \{x_2, x_3, x_6\}$ and $X'' = \{x_1, x_4, x_5, x_7\}$, the cut $\{a, b, e, g, i\}$.

The first is a proper cut set, the second is not (for it obviously contains the first). In the first case the subgraphs generated by X' and X'' are obviously connected, whereas in the second case the subgraph generated by X' is not connected.

The relationship between the cut sets of edges and the separating edge sets of a graph can be summarized as follows:

(1) *Every cut set of edges between two nodes x_i and x_j is a separating edge set between x_i and x_j,* (although the converse is not necessarily true), and
(2) *Every proper separating edge set between two nodes x_i and x_j is a proper cut set of edges between x_i and x_j, and the converse is also true.*

These statements are proved by arguments similar to those used in Section 4.2.2 (with the term 'edge' substituted for 'arc' throughout).

4.5.3. The determination of proper separating edge sets (cf. Section 4.2.3)

Let $G = (X, E)$ be a simple graph whose edges have distinct names and let $H = (X, U)$ be the graph with the same node set as G, and which has two arcs (x_i, x_j) and (x_j, x_i) between each pair of nodes x_i and x_j which are joined by an edge on G; on H, the arcs (x_i, x_j) and (x_j, x_i) both bear the name of the corresponding edge $[x_i, x_j]$ on G. As an example, for the simple graph of Fig. 4.7, the corresponding graph H is shown in Fig. 4.8.

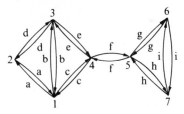

FIG. 4.8

It is evident that, if S is the set of names of the edges of an elementary chain between x_i and x_j on G, then S is also the set of names of the arcs of an elementary path from x_i to x_j on H, and the converse is also true. Thus, if we consider H to be labelled with the path algebra of Section 4.2.3, the entry a_{ij}^* of the closure of its adjacency matrix A gives the proper separating edge sets between x_i and x_j.

4.5.4 *Bridges* (*cf. Section* 4.2.4)

An edge $e = [x_i, x_j]$ of a simple graph G is called a *bridge* of G if in the graph obtained from G by removing e, the nodes x_i and x_j are not connected. In other words, an edge e is a bridge if and only if $\{e\}$ is a cut set.

As an example, the graph of Fig. 4.7 has one bridge, namely the edge $f = [x_4, x_5]$.

An edge e is called a *circuit edge* when it belongs to a circuit. We observe that *an edge e is a circuit edge if and only if it is not a bridge.* Indeed, if e is not a bridge, then the endpoints of e are joined by a simple chain which does not contain e, and by joining e to this chain we obtain a circuit. Conversely, if e is a bridge, there cannot be any chain between its endpoints which does not contain e, and therefore e does not lie on a circuit.

We say that two nodes x_i and x_j are *circuit-edge connected* if there exists a chain between x_i and x_j whose edges are all circuit edges, or if $x_i = x_j$. The relation of circuit-edge connectedness is obviously an equivalence relation on the node set X of G; the subgraphs of G generated by the equivalence classes of X are called the *leaves* of G. In other words, the leaves of G are the connected components of the partial graph of G which is obtained by removing its bridges.

As an example, the graph of Fig. 4.7 has two leaves, these being the subgraphs generated by the node sets $\{x_1, x_2, x_3, x_4\}$ and $\{x_5, x_6, x_7\}$. Again, the graph of Fig. 4.9(a) has five leaves (which are indicated by broken lines).

It follows from the above definitions that leaves have the following properties:

(1) *Every leaf of a graph is 'circuit-closed'* in the sense that if a circuit C of a graph has any node in common with a leaf L, then all the nodes and edges of C belong to L.

(2) *A leaf is connected, and does not contain any bridges*: indeed, let G be any simple graph, and let $\tilde{G}$ be the partial graph of G obtained by removing all its bridges. Since the removal of a bridge does not destroy any circuits, every circuit edge of G is also a

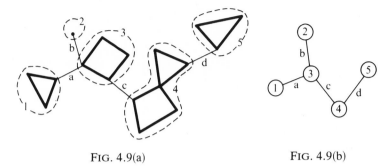

FIG. 4.9(a)   FIG. 4.9(b)

circuit edge in $\tilde{G}$, which implies that $\tilde{G}$ does not contain any bridges.

(3) *There can be at most one bridge joining two leaves*, for if two leaves were joined by more than one bridge, these bridges would become circuit edges.

Leaf graphs. Let $G = (X, E)$ be a simple graph, and let $X_1, X_2, \ldots, X_k$ be the node sets of its leaves. Then the condensation of G induced by the partition $\{X_1, X_2, \ldots, X_k\}$ is called the *leaf graph G_l of G*. Thus, the nodes of G_l correspond to the leaves of G, and two nodes in G_l are joined by an edge if and only if the corresponding leaves are connected by a bridge in G. We note that, by property (3) above, there is a one-to-one correspondence between the bridges of G and the edges of G_l. We observe also that G_l does not contain any circuits, for a circuit in G_l would represent a circuit in G passing through several leaves.

For the graph of Fig. 4.9(a) the leaf graph is shown in Fig. 4.9(b).

An algebraic method of finding bridges and leaves. Let G be a simple graph whose edges have distinct names, and let H be the corresponding directed graph, as defined in Section 4.5.3. Then, if H is considered to be labelled with the path algebra of Section 4.2.4, each entry a_{ij}^* of the closure of its adjacency matrix A is the set of names of all the bridges between x_i and x_j. Clearly, the matrix A^* is symmetrical. When two nodes x_i and x_j belong to the same leaf we have $a_{ij}^* = a_{ji}^* = \phi$, and the ith and jth rows (and columns) of A^* are identical.

Example 4.12. For the graph of Fig. 4.7,

$$A = \begin{bmatrix} \Omega & \{a\} & \{b\} & \{c\} & \Omega & \Omega & \Omega \\ \{a\} & \Omega & \{d\} & \Omega & \Omega & \Omega & \Omega \\ \{b\} & \{d\} & \Omega & \{e\} & \Omega & \Omega & \Omega \\ \{c\} & \Omega & \{e\} & \Omega & \{f\} & \Omega & \Omega \\ \Omega & \Omega & \Omega & \{f\} & \Omega & \{g\} & \{h\} \\ \Omega & \Omega & \Omega & \Omega & \{g\} & \Omega & \{i\} \\ \Omega & \Omega & \Omega & \Omega & \{h\} & \{i\} & \Omega \end{bmatrix},$$

$$A^* = \left[\begin{array}{cccc|ccc} \phi & \phi & \phi & \phi & \{f\} & \{f\} & \{f\} \\ \phi & \phi & \phi & \phi & \{f\} & \{f\} & \{f\} \\ \phi & \phi & \phi & \phi & \{f\} & \{f\} & \{f\} \\ \phi & \phi & \phi & \phi & \{f\} & \{f\} & \{f\} \\ \hline \{f\} & \{f\} & \{f\} & \{f\} & \phi & \phi & \phi \\ \{f\} & \{f\} & \{f\} & \{f\} & \phi & \phi & \phi \\ \{f\} & \{f\} & \{f\} & \{f\} & \phi & \phi & \phi \end{array} \right].$$

Another method of finding bridges and leaves, through the construction of a leaf graph, is given in Exercise 4.7.

4.6. Spanning trees in simple graphs

4.6.1. *Free trees*

A simple graph is called a *free tree* if it is connected and has no circuits. For instance, the graphs in Fig. 4.10 are all free trees.†

The following theorem lists some properties of free trees.

Let G be a simple graph; the following properties are equivalent, for characterizing a free tree:

(1) *G is connected and contains no circuits;*
(2) *G is connected and every edge of G is a bridge;*
(3) *Every pair of nodes of G is joined by precisely one chain;*
(4) *G contains no circuits, but the addition of any new edge creates a circuit.*

---

† Graphs of this kind are often simply called 'trees'. However, in Computer science and other applications areas the term 'tree' is generally used for graphs of the kind described in Section 2.6, which have a root, and branches directed away from the root. The nomenclature used here is that of Knuth (1968).

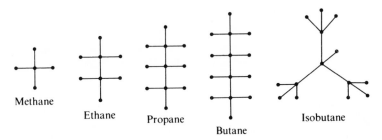

FIG. 4.10. Graphs of the saturated hydrocarbons C_2H_{2n+2} which have up to four carbon atoms. (Nodes of degree four represent carbon atoms, nodes of degree one are hydrogen atoms.)

The theorem is proved as follows:

(*1*) *implies* (*2*). Indeed, it was proved in the previous section that in a simple graph, any edge which does not belong to a circuit is a bridge. Since G has no circuits, all its edges are bridges.

(*2*) *implies* (*3*). Since G is connected, every pair of nodes is joined by at least one chain. Also, a given pair of nodes cannot be joined by two chains, for otherwise the removal of an edge which belonged only to one of them would not disconnect the graph.

(*3*) *implies* (*4*). If G contained a circuit, then any two nodes in the circuit would be joined by two chains. The addition of any new edge e creates a circuit because in G, the endpoints of e are already joined by a chain.

(*4*) *implies* (*1*). G is connected because if G had two nodes which were not connected, it would be possible to add to G an edge between these nodes, without creating a circuit.

From the characterization (2) above, it follows that *every free tree with n nodes has precisely n − 1 edges*. Indeed, if an n-node graph G is a free tree then the removal of any edge divides G into two connected components, which are both free trees. Then, the removal of a second edge divides G into three free trees, and so on. When we have removed all its edges, the graph G comprises n free trees (which are all isolated nodes). The number of edges removed in this process is obviously $n − 1$.

4.6.2. Spanning trees

Let G be a simple graph. Then any partial graph of G which is a free tree is called a *spanning tree* of G. As an example, Fig. 4.11 shows two of the spanning trees of the graph of Fig. 4.7.

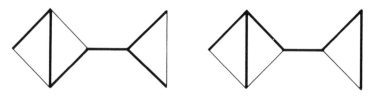

FIG. 4.11

Clearly, a graph must be connected to have a spanning tree. Also, any connected graph has at least one spanning tree: indeed, for any connected graph G, we can always construct a spanning tree as follows. If all the edges of G are bridges, then (by property (2) above) G is a free tree; otherwise, if G has an edge which is not a bridge, it can be removed without disconnecting G, and by the successive removal of such edges we eventually obtain a spanning tree.

It is evident that a spanning tree represents a minimal collection of edges which preserves the connectedness of a graph. This concept is in a sense complementary to that of a proper cut set of edges (which is a minimal collection of edges whose removal disconnects some nodes from others). These notions are related precisely by the following theorem:

In a connected graph, every cut set of edges has at least one edge in common with every spanning tree.

To prove the theorem, let C be a cut set of edges of a graph G, and let T be a spanning tree of G. Then, if C did not contain at least one edge from T, the removal of C from G would not separate G into two or more components.

4.6.3. Shortest spanning trees

Let us consider the following problem. In an electrical network there are n terminals, in fixed positions, which must all be electrically connected together by wires. Which pairs of terminals should be joined, if the total length of wire used is to be as small as possible?

This problem can be presented in graph-theoretic terms, as follows. Let $G = (X, E)$ be a simple graph in which every edge e is assigned a real number $l(e)$, called its *length*. It is required to find a connected partial graph $H = (X, F)$ of G whose *total length*

$$l(H) = \sum_{e \in F} l(e)$$

is as small as possible. It is evident that H must be a spanning tree of G: we describe such a graph as a *shortest* spanning tree.

The construction of a shortest spanning tree. A shortest spanning tree can be 'grown' from an arbitrarily chosen node $x^{(0)}$, one edge at a time, by constructing a sequence of graphs $H^{(k)} = (X^{(k)}, F^{(k)})$, $(k = 0, 1, \ldots, n-1)$, as follows.

Step 1 Let $x^{(0)}$ be any node of G, and let $X^{(0)} = \{x^{(0)}\}$, $F^{(0)} = \phi$.

Step 2 For $k = 1, 2, \ldots, n-1$, form $X^{(k)}$ and $F^{(k)}$ as follows: let $C^{(k)}$ be the cut set of edges which have exactly one endpoint in $X^{(k-1)}$, let $e^{(k)}$ be a shortest edge of $C^{(k)}$, and let $x^{(k)}$ be the endpoint of $e^{(k)}$ which does not belong to $X^{(k-1)}$. Then set

$$X^{(k)} = X^{(k-1)} \cup \{x^{(k)}\}, \qquad F^{(k)} = F^{(k-1)} \cup \{e^{(k)}\}. \qquad (4.7)$$

On termination, the graph $H^{(n-1)} = (X^{(n-1)}, F^{(n-1)})$ is a shortest spanning tree of G. To demonstrate this, it suffices to prove the following.

Each graph $H^{(k)} = (X^{(k)}, F^{(k)})$ constructed by the algorithm is a $(k+1)$-node connected subgraph of a shortest spanning tree on G. The proof is by induction. For $k = 0$ the statement is evidently true. Let us therefore suppose that it holds for $k = r - 1$, and demonstrate its validity for $k = r$. It is clear from (4.7) that since $H^{(r-1)}$ is an r-node connected graph, $H^{(r)}$ has $r+1$ nodes, and is connected. Now let H be any shortest spanning tree of G which contains all the edges of $H^{(r-1)}$. If H also contains $e^{(r)}$, then $H^{(r)}$ is a connected subgraph of H. Alternatively, let us suppose that H does not contain $e^{(r)}$. Then if $e^{(r)}$ is added to H, a circuit is created, composed of $e^{(r)}$ together with the chain in H between the endpoints of $e^{(r)}$. Now this chain must contain some edge f say of $C^{(r)}$. If f is removed from H we obtain another spanning tree of G, and since $l(e^{(r)}) \le l(f)$, this spanning tree is another shortest spanning tree of

G. Hence there exists a shortest tree containing all the edges of $H^{(r)}$, which completes the proof.

It will be noted that at the *k*th step, the chosen edge is always a *shortest* edge among the edges between $x^{(k)}$ and the nodes of $X^{(k-1)}$. It follows that at each stage, after $x^{(k)}$ has been selected, if any node *y* not in $X^{(k)}$ has edges joining it both to $x^{(k)}$ and to some other node in $X^{(k)}$, the longer of these two edges can immediately be deleted from *G*. This technique substantially reduces the number of comparisons of edge lengths required (see Exercise 4.6).

Example 4.13. The construction of a shortest spanning tree is demonstrated in Fig. 4.12. At each stage, the bold lines are the edges of the graph $H^{(k-1)}$, the closed curve indicates the cut set of edges $C^{(k)}$, and the broken line indicates the next edge $e^{(k)}$ to be assigned to the spanning tree.

For the case where we require a single spanning tree on a graph $G = (X, E)$ whose edge lengths can be considered to be equal, it is convenient to recast the above algorithm in the following form. (In

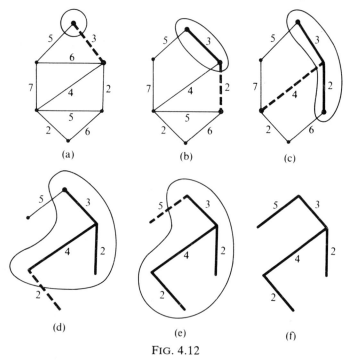

(a)   (b)   (c)

(d)   (e)   (f)

FIG. 4.12

this version of the algorithm we again 'grow' a spanning tree from an arbitrarily chosen node x by successively adding edges and nodes to a connected subgraph $H = (\tilde{X}, \tilde{E})$ of a spanning tree of G. However, instead of adding edges one at a time, each step of the algorithm joins to H all the neighbours of some node y in H.)

Step 1   Let x be any node of G, and let $\tilde{X} = \{x\}$, $\tilde{E} = \phi$, $Y = \{x\}$.

Step 2   Let y be any node in Y, and let Z be the set of all neighbours of y which do not belong to X:

$$Z = \Gamma(y) - \tilde{X}.$$

Then modify the sets $\tilde{X}$, $\tilde{E}$, and Y as follows:

$$\tilde{X} \leftarrow \tilde{X} \cup Z,$$
$$\tilde{E} \leftarrow \tilde{E} \cup \{[y, z] \mid z \in Z\},$$
$$Y \leftarrow (Y - \{y\}) \cup Z.$$

Step 3   If $Y = \phi$ then halt; otherwise return to *Step 2*.

On termination, the graph $H = (\tilde{X}, \tilde{E})$ is a spanning tree of G.

4.6.4. *Determination of the spanning trees of a graph*

It is possible to obtain all the spanning trees of a graph by a backtrack programming method of the kind described in Section 2.7. To present the method it is convenient to consider a more general problem—that of finding the spanning trees of a given *multigraph*. (Our definition of a spanning tree in Section 4.6.2 still applies when G is a multigraph; we note in particular that even for a multigraph, a spanning tree cannot have more than one edge joining the same pair of nodes.) Our reason for considering this more general problem is that in searching for spanning trees, even on a simple graph, we may create sub-problems which involve multigraphs.

Now let G be a connected multigraph, and let S be the set of all spanning trees of G. Let e be any edge of G, let S_e be the set of spanning trees of G which contain e, and let $S_{\bar{e}}$ be the set of spanning trees of G which do not contain e. Clearly,

$$S = S_e \cup S_{\bar{e}} \quad \text{and} \quad S_e \cap S_{\bar{e}} = \phi,$$

so if we can construct the two sets S_e and $S_{\bar{e}}$ separately we shall obtain all the required spanning trees, without duplications. Now it will be observed that:

(i) Each spanning tree in S_e comprises the edge e, together with the edges of a spanning tree of the multigraph G_e obtained by *contracting* the edge e, that is, by coalescing its endpoints and removing any loops created (see Fig. 4.13); and conversely, the edges of any spanning tree in G_e, together with the edge e, form a spanning tree in S_e.

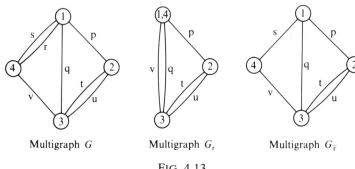

Multigraph G Multigraph G_r Multigraph $G_{\bar{r}}$

FIG. 4.13

(ii) The spanning trees in $S_{\bar{e}}$ are the spanning trees of the multigraph $G_{\bar{e}}$ obtained from G by removing the edge e. (Note that if e is a bridge, $S_{\bar{e}}$ is an empty set; to determine whether or not e is a bridge it is convenient here to establish whether one endpoint of e is accessible from the other in $G_{\bar{e}}$, by the labelling algorithm of Fig. 2.19.)

Thus we can decompose the problem of finding the spanning trees of G into two sub-problems, each involving the determination of the spanning trees of a multigraph which has less edges than G. By repeated decompositions of this kind, we eventually obtain all the required spanning trees.

4.7. Node separation on simple graphs

In this section we consider ways of destroying all the chains between two points on a simple graph, by removing some of its nodes.

4.7.1. Articulation sets (cf. Section 4.4.1)

Let x_i and x_j be any two nodes (which need not be distinct) on a connected simple graph $G = (X, E)$. Then an *articulation set between x_i and x_j* is a node set W, not containing x_i or x_j, such that every chain between x_i and x_j traverses at least one node in W. When W is an articulation set between x_i and x_j, and none of its proper subsets have this property, we say that W is a *proper articulation set* between x_i and x_j.

It is evident that when the nodes of an articulation set are removed from a graph, the graph becomes disconnected.

Example 4.14. On the graph of Fig. 4.7, the proper articulation sets between the nodes x_2 and x_6 are $\{x_1, x_3\}$, $\{x_4\}$ and $\{x_5\}$.

The determination of articulation sets. Let $G = (X, E)$ be a simple graph and let $H = (X, U)$ be the graph with the same node set X as G, and which has two arcs (x_i, x_j) and (x_j, x_i) between each pair of nodes x_i and x_j which are joined by an edge of G. Then for any two nodes x_k and x_l, each proper articulation set between x_k and x_l on G is a proper (x_k, x_l)-separating node set on H, and the converse is also true. It is therefore possible to find the proper articulation sets of G by applying the technique of Section 4.4.1 to the graph H.

4.7.2. Articulation nodes (cf. Section 4.4.2)

A node x_k in a connected simple graph G is an *articulation node* of G if the graph obtained from G by removing x_k and all edges incident to x_k is not connected. In other words, a node x_k is an articulation node if and only if $\{x_k\}$ is an articulation set between some pair of nodes x_i and x_j.

As an illustration, the graph of Fig. 4.7 has two articulation nodes, x_4 and x_5.

Bi-connected graphs. Two edges e_1 and e_2 of a graph G are said to be *strongly circuit-connected* if G has an elementary circuit which contains both e_1 and e_2, or if $e_1 = e_2$. It is evident that the relation of strong circuit-connectedness between edges is reflexive, and symmetric. We can also show it to be transitive, by proving the following theorem:

Let e_1, e_2, and e_3 be three edges of a graph G. If e_1 is strongly circuit-connected to e_2, and e_2 is strongly circuit-connected to e_3, then e_1 is strongly circuit-connected to e_3.

To prove the theorem, let α be any elementary circuit which contains e_1 and e_2, and let β be any elementary circuit containing e_2 and e_3. Since e_2 belongs to both α and β, these circuits have at least two nodes in common (see Fig. 4.14). Following α in both directions from e_1, let x and y be the first two nodes on α which lie also on β, and let α' and β' be the segments of α and β which join x to y and which contain e_1 and e_3 respectively. Since the segments α' and β' have no nodes in common other than their endpoints, these two segments form an elementary circuit. Thus e_1 and e_3 lie on an elementary circuit, as required.

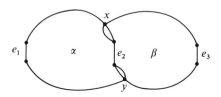

FIG. 4.14

From this result we conclude that the relation of strong circuit-connectedness is an equivalence relation on the edge set of G.

Now we can relate this notion of connectedness to that of an articulation node:

Let G be a simple graph with at least three nodes. Then the following statements are equivalent:

(1) *G is connected, and has no articulation nodes*

(2) *G is connected, and any two edges of G are strongly circuit-connected.*

(3) *Any two nodes of G lie on a common elementary circuit.*

The proof is as follows:

(1) implies (2). First we show that when G has no articulation nodes, any two edges with a common endpoint are strongly circuit-connected: indeed, if g contains two edges $e = [x_i, x_k]$ and $f = [x_k, x_j]$, and x_k is not an articulation node, then x_i and x_j must be joined by a chain which does not traverse x_k. It follows that x_i and x_j are joined by an elementary chain which does not traverse x_k, and the edges of this chain, together with e and f, form an elementary circuit. Hence e and f are strongly circuit-connected.

Now let e and f be any two edges in G. Since G is connected, G contains a chain C whose first edge is e and whose last edge is f. By the above argument, each edge of C is strongly circuit-connected to its succeeding edge; it follows, by the transitivity of the circuit-connectedness relation, that e is strongly circuit-connected to f.

(2) implies (3). Since G is connected, every node is the endpoint of an edge, and since any two edges lie on a common elementary circuit, any two nodes lie on a common elementary circuit.

(3) implies (1). Any two nodes are connected, therefore G is connected. Also, for any two nodes x_i and x_j, there does not exist a third node x_k which lies on every chain between x_i and x_j, and therefore G has no articulation nodes.

A graph with at least three nodes, which has these properties, is said to be *bi-connected*.

The blocks of a graph. In section 4.5.4 we introduced an equivalence relation on the node set of a graph—that of circuit-edge connectedness—and we saw that it defined a 'decomposition' of a graph into 'leaves', these being the connected subgraphs obtained by cutting all the bridges.

Earlier in this section we introduced the relation of strong circuit-connectedness, and we showed that this relation is an equivalence relation on the edge set of a graph. We shall now demonstrate that this relation leads to another way of decomposing a graph, in which this time we 'split' the articulation nodes.

For a graph $G = (X, E)$, let $P = \{E_1, E_2, \ldots, E_k\}$ be the partition of the edge set E induced by the strong circuit-connectedness relation, and for each block E_i in P, let X_i be the set of nodes which are endpoints of edges in E_i. Each partial graph $B_i = (X_i, E_i)$ of G is called a *block* of G.

Example 4.15. For the graph of Fig. 4.7, the equivalence classes of strongly circuit-connected edges are

$$\{a, b, c, d, e\}, \{f\}, \{g, h, i\}.$$

The blocks associated with these equivalence classes are shown in Fig. 4.15.

The blocks of a graph have the following properties:

(1) *Every block is 'circuit-closed'* in the sense that if an elementary circuit C of G has any edge in common with a block B, then all the edges of C appear in B.

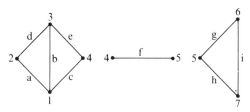

FIG. 4.15

(2) *A block is connected, and does not contain any articulation nodes.* In the case where a block B comprises exactly one edge and its endpoints, this statement is obviously true. Now let us suppose that a block B has more than one edge, and let e and f be any two edges in B. Then in G, the edges e and f belong to a common elementary circuit, and by property (1) above all the edges of this circuit appear in B. It follows that in B, every pair of edges is strongly circuit-connected. Furthermore, since every node in B is the endpoint of an edge, B must be connected. Thus B is a bi-connected graph, as required.

(3) *Every block of a graph G is a subgraph of G.* To prove this it suffices to show that if two nodes of a block B are joined by an edge e in G then the edge e belongs to B. If B has only two nodes this is evidently true. Alternatively, if B has more than two nodes then by (2) above B is bi-connected, which implies that in B any pair of nodes x_i and x_j lie on a common elementary circuit, C say. If G contains an edge $e = [x_i, x_j]$, this edge forms an elementary circuit with one of the segments of C which join x_i to x_j, which implies that e belongs to B.

(4) *Two different blocks have at most one node in common.* If two blocks had two nodes in common then, since each block is connected, there would exist an elementary circuit with at least one edge in each block, which is impossible.

(5) *A node of G is an articulation node of G if and only if it is common to two distinct blocks of G.* Let x_k be a node common to two blocks B_1 and B_2 of G, and let $e = [x_i, x_k]$ and $f = [x_j, x_k]$ be any edges incident to x_k, in B_1 and B_2 respectively. Since e and f do not lie on a common elementary circuit, every chain between x_i and x_j traverses x_k, which implies that x_k is an articulation node. Conversely, if x_k is an articulation node, there exist two edges incident

to x_k which do not lie on an elementary circuit. These two edges belong to different blocks, and x_k belongs to the node set of both blocks.

Block graphs. The 'block structure' of a simple graph $G = (X, E)$ can be portrayed by constructing its block graph G_b, whose nodes correspond to the blocks of G, and in which two nodes are joined by an edge if the corresponding blocks of G have a common articulation node of G.

As an illustration, the graph of Fig. 4.16(a) has the block decomposition depicted by Fig. 4.16(b); the corresponding block graph is given in Fig. 4.16(c).

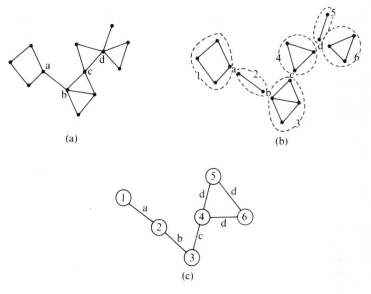

(a)

(b)

(c)

FIG. 4.16

It is clear that within a block graph G_b, each block is a complete graph, which corresponds to an articulation node of G.

Algebraic methods of finding the articulation nodes and blocks of a graph. Let $G = (X, E)$ be a simple graph and let $H = (X, U)$ be the corresponding symmetric graph, having two arcs (x_i, x_j) and (x_j, x_i) in place of each edge $[x_i, x_j]$ of G. Then for any two nodes $x_k, x_l \in X$, each articulation node between x_k and x_l on G is an (x_k, x_l)-separating node on H, and the converse is also true. The articulation nodes of G can therefore be obtained by applying the technique of Section 4.4.2 to the graph H.

An alternative approach is as follows. Given any simple graph $G = (X, E)$, we define the *interchange graph I* of G as the graph whose nodes represent the edges of G, and in which two nodes e_i and e_j are joined by an edge $[e_i, e_j]$ if in G the edges e_i and e_j have a common endpoint. As an illustration, Fig. 4.17 shows the interchange graph of Fig. 4.7.

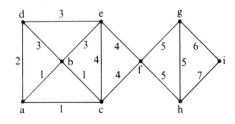

FIG. 4.17

Now let us suppose that each edge $[e_i, e_j]$ on the interchange graph I is labelled with the index of the common endpoint of the edges e_i and e_j on G (cf. Figs. 4.7 and 4.17) and let H be the corresponding symmetric graph obtained by replacing each edge $[x_i, x_j]$ by a pair of arcs (x_i, x_j) and (x_j, x_i), both these bearing the same name as the edge which they replace. Then we may consider H as a graph labelled with the path algebra of Section 4.2.4; if A is the adjacency matrix of H, it follows from (4.6) that the entries of A^* are as follows:

(i) if on G the edges e_i and e_j belong to different connected components then $a_{ij}^* = \Omega$;

(ii) if on G the edges e_i and e_j belong to the same connected component then a_{ij}^* is the set of all articulation nodes between the block containing e_i and the block containing e_j; in particular, if e_i and e_j belong to the same block then $a_{ij}^* = \phi$.

Example 4.16. For the interchange graph of Fig. 4.17, the A-matrix and A^*-matrix are given in Fig. 4.18. The entries of A^* define the blocks and articulation nodes of G (see Fig. 4.15).

	a	b	c	d	e	f	g	h	i
a	Ω	{1}	{1}	{2}	Ω	Ω	Ω	Ω	Ω
b	{1}	Ω	{1}	{3}	{3}	Ω	Ω	Ω	Ω
c	{1}	{1}	Ω	Ω	{4}	{4}	Ω	Ω	Ω
d	{2}	{3}	Ω	Ω	{3}	Ω	Ω	Ω	Ω
e	Ω	{3}	{4}	{3}	Ω	{4}	Ω	Ω	Ω
f	Ω	Ω	{4}	Ω	{4}	Ω	{5}	{5}	Ω
g	Ω	Ω	Ω	Ω	Ω	{5}	Ω	{5}	{6}
h	Ω	Ω	Ω	Ω	Ω	{5}	{5}	Ω	{7}
i	Ω	Ω	Ω	Ω	Ω	Ω	{6}	{7}	Ω

(a) A-matrix

	a	b	c	d	e	f	g	h	i
a	φ	φ	φ	φ	φ	{4}	{4, 5}	{4, 5}	{4, 5}
b	φ	φ	φ	φ	φ	{4}	{4, 5}	{4, 5}	{4, 5}
c	φ	φ	φ	φ	φ	{4}	{4, 5}	{4, 5}	{4, 5}
d	φ	φ	φ	φ	φ	{4}	{4, 5}	{4, 5}	{4, 5}
e	φ	φ	φ	φ	φ	{4}	{4, 5}	{4, 5}	{4, 5}
f	{4}	{4}	{4}	{4}	{4}	φ	{5}	{5}	{5}
g	{4, 5}	{4, 5}	{4, 5}	{4, 5}	{4, 5}	{5}	φ	φ	φ
h	{4, 5}	{4, 5}	{4, 5}	{4, 5}	{4, 5}	{5}	φ	φ	φ
i	{4, 5}	{4, 5}	{4, 5}	{4, 5}	{4, 5}	{5}	φ	φ	φ

(b) A^*-matrix

Fig. 4.18

Exercises

4.1. Prove that every Hamiltonian cycle on a graph G contains all the basic arcs of G.

4.2. Using the path algebra of Section 4.2.4, construct the weak closure matrix A of the graph of Fig. 4.19, and hence find its basic arcs.

4.3. Let x_s be any node on a graph $G = (X, U)$. We say that on G, a node x_i is a *dominator* of another node x_j (relative to x_s) if x_i is an (x_s, x_j)-separating node, or $x_i = x_s$. A node x_i is an *immediate* (or *direct*) *dominator* of another node x_j (relative to x_s) if

   (a) x_i dominates x_j, and
   (b) every other dominator of x_j also dominates x_i.

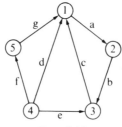

FIG. 4.19

(i) Prove that if all the nodes of G are accessible from x_s, the graph of the immediate dominance relation on X is a tree rooted at x_s, and show that this tree contains a path from a node x_i to a node x_j if and only if x_i dominates x_j. (This tree is called the *domination tree of G, relative to x_s*.)
(ii) Prove that if G is a tree rooted at x_s, then the domination tree of G relative to x_s is identical to G.
(iii) Let G be any graph in which all nodes are accessible from x_s, and let T be its domination tree (relative to x_s). Let $\tilde{G}$ be the graph obtained by adding some arc (x_i, x_j) to G, and let $\tilde{T}$ be the new domination tree. Formulate rules for constructing $\tilde{T}$ from T, in each of the three following cases:

    (a) T contains a path from x_i to x_j;
    (b) T contains a path from x_j to x_i;
    (c) T does not contain any paths between x_i and x_j.

(The results of (ii) and (iii) suggest a method of constructing a domination tree. First we choose a partial graph H of G which is a tree rooted at x_s; we then obtain the domination tree of G by making a succession of modifications to H, to take account of the arcs of G which were not assigned to H initially.)

4.4. Prove that on a simple graph, every circuit has an even number of edges in common with every cut set.

4.5. Let T_1 and T_2 be two spanning trees of a simple graph G. Prove that if a is any edge in T_1, then there exists an edge b in T_2 such that the graph obtained from T_1 by replacing a by b is a spanning tree of G.

4.6. Show that, in the algorithm of Section 4.6.3 for finding a shortest spanning tree, the number of comparisons of edge lengths is not greater than $(n-1)(n-2)$.

4.7. (i) Let G be a free tree. Explain how the nodes and edges of the leaf graph G_l of G are related to the nodes and edges of G.

(ii) Let G be a connected simple graph, and let G_l be its leaf **graph.** Also, let $\tilde{G}$ be the graph obtained by adding some edge e to G and let $\tilde{G}_l$ be the leaf graph of $\tilde{G}$. Explain how $\tilde{G}_l$ can be derived from G_l in the cases where (a) the endpoints of e belong to the same leaf in G, and (b) the endpoints of e belong to different leaves in G.

(These rules suggest a simple method of finding the leaves and bridges of a connected graph G. First we construct a spanning tree T of G, for instance by the algorithm described in Section 4.6.3; we then derive the leaf graph of G by a succession of modifications of the leaf graph of T, to take account of the edges of G which do not appear in T.)

Additional notes and bibliography

The notions of separating arc and node sets on directed graphs are discussed at length by Harary, Norman, and Cartwright (1965).

Martelli (1976) discusses the application of the path algebra of Section 4.2.3 to the determination of all the proper cut sets of arcs, between each pair of nodes of a graph.

It will be noted that the number of proper cut sets between two nodes of a graph can be very large, and their computation may not be practically feasible. However, if one only requires one cut set of minimum cardinality, this can be found conveniently by a network flow method, which will be presented in Chapter 6. (The network flow method, which is a polynomial algorithm, can also be used for instance when the arcs of a graph have numerical weights, to find a cut set of minimum total weight between two specified points.)

The path algebra for finding basic arcs (Section 4.2.4) is apparently new. However, it is very similar to the 'distributive monotone framework' given by Hecht (1977) for finding dominator nodes, and our formulations of the problem of finding dominators are essentially the same. The 'round robin' algorithm used by Hecht to solve this and other program data flow problems can be seen to be an extension of the Gauss–Seidel method.

Tarjan (1974a) gives an efficient algorithm for finding immediate dominators, based on the principles set out in Exercise 4.3. Other techniques for finding dominators are described by Lowry and Medlock (1969), and Purdom and Moore (1972).

A general algorithm for constructing an arbitrary basis graph of a graph is described by Moyles and Thompson (1969).

The notion of a basis graph can be extended to graphs labelled with certain path algebras. For instance, given a graph G whose arcs have lengths, we may wish to find a partial graph H of G such that (i) the distance between any two points is the same on G and H, and (ii) the number of arcs of H is minimal. An extension of the notion of a basis graph to include partial graphs of this kind is presented by Robert (1971).

The algebraic structure for finding cut sets of edges is discussed by Hulme (1975). For an alternative backtrack programming method of finding these cut sets, and applications to system reliability studies, see Jensen and Bellmore (1969).

Different algorithms for finding bridges and blocks are given by Paton (1971), Corneil (1971), Tarjan (1972) and Hopcroft and Tarjan (1973). See also Tarjan (1974b).

The method described in Section 4.6.3 for constructing shortest spanning trees, which is due to Prim (1957), is a particular example of a class of methods first proposed by Kruskal (1956) (see Kruskal's 'Construction B'). For a FORTRAN version of Prim's algorithm, see Whitney (1972). Yao (1975) gives an alternative method which has a lower time bound but which is more intricate. See also Kershenbaum and Van Slyke (1972), Cheriton and Tarjan (1976), and Gabow (1977).

A backtrack programming method of finding all the spanning trees of a graph, based on the strategy of Section 4.6.4, has been outlined by Minty (1965), and implemented in ALGOL by McIlroy (1969). For an analysis of its complexity see Read and Tarjan (1975). The spanning trees of a graph can also be found using Wang's algebra of networks (Duffin 1959; Chen 1971). See also Trent (1954).

5 Independent sets, dominating sets, and colorations

5.1. Introduction

THE TOPICS discussed in this chapter all relate to simple graphs. In particular, our purpose is to characterize certain subsets of nodes on a simple graph (the independent node sets, cliques, and dominating sets), and also certain subsets of edges (the matchings of a graph), and to demonstrate their practical relevance. We shall then present ways of partitioning the sets of nodes and edges, which may be interpreted as 'colorations'; these also find practical applications, for instance in the construction of time-tables.

5.2. Independent sets

5.2.1. Independent node sets

In a simple graph $G = (X, E)$, a set of nodes $S \subseteq X$ is said to be *independent* if no two nodes in S are joined by an edge. Clearly, every subset of an independent set is independent. A *maximal independent set* is an independent set which ceases to be independent when any node is added to it.

In the graph of Fig. 5.1, the maximal independent sets are $\{x_1, x_5\}$, $\{x_2, x_3, x_5\}$, $\{x_2, x_5, x_6\}$, $\{x_4\}$.

The largest number of nodes in an independent set of a graph G is called the *independence number* of G and is denoted by $\alpha(G)$. For the graph of Fig. 5.1, $\alpha(G) = 3$.

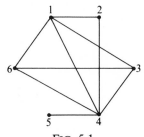

FIG. 5.1

Example 5.1. *Error-detecting and error-correcting codes.* Let $X = \{x_1, x_2, \ldots, x_n\}$ be the set of basic signals which can be transmitted through a digital communication channel. (For instance, X could be a set of binary words such as are used to represent alphanumeric characters in a computer system.) Because of electrical noise and distortion, some transmitted signals may be misinterpreted on reception. In general, for each pair of signals x_i and x_j, the reception of x_i as x_j occurs with a different probability, but for practical purposes it is sometimes adequate to consider the definite case where, for any pair of signals x_i and x_j, the reception of x_i as x_j either can or cannot occur: in this case, the possible communication errors can be represented conveniently by a *signal-relation graph* $G = (X, E)$ whose nodes correspond to the basic signals, and where two nodes x_i and x_j are joined by an edge if either of these signals can be received as the other (see Fig. 5.2).

Signal transmitted	Signal received
a	a or b
b	b or c
c	c or d
d	d or e
e	b or e or f
f	a or f

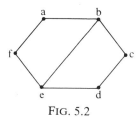

FIG. 5.2

Now it is possible to *detect* all errors in received signals, if we restrict the transmitted signals to a subset of X which is independent on the signal-relation graph G. Indeed, let us suppose that the set S of transmitted signals is independent; then if a received signal belongs to S it is correct (since no signal in S can be transformed into any other signal in S), whereas if a received signal does not belong to S, it is evidently incorrect. On the other hand, if the set of transmitted signals is not independent, some pair of transmitted signals can be confused by the receiver.

In the example of Fig. 5.2 the largest independent sets, which determine the largest sets of signals which can be used with detection of all errors, are $\{a, c, e\}$ and $\{b, d, f\}$.

It is also possible to *correct* errors in received signals, as follows. Let us construct another signal-relation graph H, whose nodes correspond to the signals as before, but in which two nodes x_i and x_j are now joined by an edge if and only if the transmission of x_i and x_j can result in the same received signal (which need not be x_i or x_j). For instance, the graph H for our example above is shown in Fig. 5.3.

Now let us suppose that the set S of transmitted signals is independent on H. Then, if a received signal x_i belongs to S it must be correct (since no

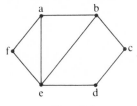

FIG. 5.3

signal in S can be transformed into any other signal in S); if x_i does not belong to S then x_i is incorrect, but since there is only one signal in S which can be received as x_i, the error can be corrected. On the other hand, if S is not independent on H, there exist two different signals in S which can be received as the same signal; in this case it is obviously impossible to determine at the receiver which signal was transmitted. Thus, the independent sets on H determine the sets of transmitted signals for which all errors can be corrected by the receiver.

For the above example, the set of signals {b, d, f} is the largest for which all errors can be corrected. (Error correction is not always possible for the set {a, c, e}, since the transmission of 'a' and 'e' can both result in a 'b').

5.2.2. Cliques

A set C of nodes in a simple graph G is called a *clique* of G if the subgraph of G generated by C is complete, that is, if every pair of nodes in C is connected by an edge. A clique which is not a subset of a larger clique is said to be *maximal*.

It is clear that a set of nodes C is a clique of a simple graph G if and only if C is an independent set on the complement $\bar{G}$ of G.

In the graph of Fig. 5.1, the maximal cliques are $\{x_1, x_3, x_4, x_6\}$, $\{x_1, x_2, x_4\}, \{x_4, x_5\}$. The complete subgraph generated by the first of these cliques is indicated by bold lines in Fig. 5.4. The nodes of the corresponding maximal independent set, on the complementary graph, are indicated by squares in Fig. 5.5.

The concept of a clique is important in taxonomy (Augustson and Minker 1970) and in the design of sequential logic networks (Paull and Unger 1959).

An algorithm for constructing maximal cliques. It is possible to find all the maximal cliques of a graph by a tree-search method of the type described in Section 2.7. To present the method, it will be convenient to consider the following more general problem:

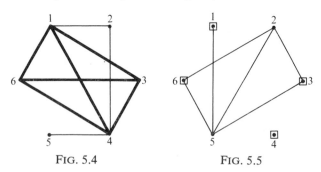

FIG. 5.4   FIG. 5.5

Given a simple graph $G = (X, E)$, together with a specified sub-graph $\tilde{G} = (\tilde{X}, \tilde{E})$ of G, and a specified node set $N \subseteq X - \tilde{X}$, find all the maximal cliques of $\tilde{G}$ which are not contained in the set $\Gamma(x_i)$ of neighbours on G of any node $x_i \in N$.

It will be noted that our original problem of finding all the maximal cliques of a graph G is the particular form of this new problem in which $\tilde{G} = G$ (which implies $N = \phi$).

To solve the general problem, let $\tilde{S}$ denote the set of all maximal cliques of $\tilde{G}$, and let S be the set of required cliques; then we may write

$$S = \{C \in \tilde{S} \mid C \not\subseteq \Gamma(x_i), \text{ for all } x_i \in N\}. \qquad (5.1)$$

Now this set S can be constructed by the following specialization process.

Step 1   First we determine whether or not the node set $\tilde{X}$ of $\tilde{G}$ satisfies the condition

$$\tilde{X} \subseteq \Gamma(x_i) \quad \text{for some } x_i \in N. \qquad (5.2)$$

If this condition holds, then for every clique $C \in \tilde{S}$ we have

$$C \subseteq \Gamma(x_i) \quad \text{for some } x_i \in N, \qquad (5.3)$$

which implies (by (5.1)) that the required set S is null, and the problem is solved; otherwise we proceed to *Step 2*.

Step 2   If the subgraph $\tilde{G} = (\tilde{X}, \tilde{E})$ is complete then $\tilde{S} = \{\tilde{X}\}$, and since the condition (5.2) does not hold it follows (by (5.1)) that $S = \{\tilde{X}\}$, so the problem is solved; otherwise we proceed to *Step 3*.

Step 3   Since the subgraph $\tilde{G}$ is not complete, it has at least one node x_k which is not adjacent to every node of $\tilde{X}$. Now let us express the required set S as the union of two disjoint sets

$$S = S_k \cup S_{\bar{k}} \qquad (5.4(a))$$

where

$$S_k = \{C \in S \mid x_k \in C\}, \qquad S_{\bar{k}} = \{C \in S \mid x_k \notin C\}. \quad (5.4(b))$$

From (5.1) and (5.4(b)), it follows that (cf. (5.1))

$$S_k = \{C \in \tilde{S}_k \mid C \nsubseteq \Gamma(x_i), \text{ for all } x_i \in N_k\}, \qquad (5.5)$$

where $\tilde{S}_k$ is the set of all maximal cliques on the subgraph $\tilde{G}_k$ of $\tilde{G}$ which contains only the node x_k and its neighbours, and $N_k = N$; whereas

$$S_{\bar{k}} = \{C \in \tilde{S}_{\bar{k}} \mid C \nsubseteq \Gamma(x_i), \text{ for all } x_i \in N_{\bar{k}}\}, \qquad (5.6)$$

where $\tilde{S}_{\bar{k}}$ is the set of all maximal cliques on the subgraph $\tilde{G}_{\bar{k}}$ of $\tilde{G}$ obtained by removing x_k, and $N_{\bar{k}} = N \cup \{x_k\}$. Thus the problem of finding the set S of (5.1) is reduced to two sub-problems, involving the separate determination of the sets S_k and $S_{\bar{k}}$ defined by (5.5) and (5.6) respectively.

It will be observed that the problems of constructing the sets S_k and $S_{\bar{k}}$ are both 'simpler' than the original problem of finding S, in that the subgraphs $\tilde{G}_k$ and $\tilde{G}_{\bar{k}}$ both contain less nodes than $\tilde{G}$. Consequently, the repeated application of the specialization process ultimately yields sub-problems which are all 'trivial' (in that for each sub-problem, either condition (5.2) holds or the subgraph $\tilde{G}$ is complete).

The use of this method, for finding the maximal cliques of the graph of Fig. 5.1, is depicted by the search tree of Fig. 5.6. Here the root node represents the original problem (for which $\tilde{G} = G$, and $N = \phi$). The labels on the arcs of the search tree indicate which nodes have been selected, in performing the specializations; at each node of the tree—which represents a sub-problem—we have given the corresponding subgraph $\tilde{G}$ and node set N.

A backtrack programming algorithm based on this specialization method has been published by Bron and Kerbosch (1973). (Their algorithm also incorporates a particular method of selecting the nodes for the successive specializations, whose aim is to minimize the total number of selections required.)

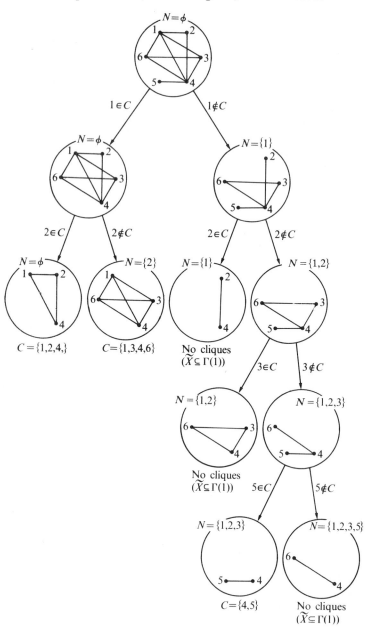

FIG. 5.6

A simplification rule. The number of specializations involved in finding all the maximal cliques of a graph can often be reduced considerably, by applying the following rule:

Let $G = (X, E)$ be a simple graph, let C be any maximal clique of G, and let x_k be any node such that $x_k \notin C$; if x_k has any neighbour x_l such that

$$\{x_l\} \cup \Gamma(x_l) \subseteq \{x_k\} \cup \Gamma(x_k) \tag{5.7}$$

then $x_l \notin C$.

Indeed, it is evident that since the clique C is maximal, and $x_k \notin C$, the clique C contains at least one node x_j which is not adjacent to x_k; but then x_j is not adjacent to any node x_l for which the condition (5.7) holds, and therefore C cannot contain any of these nodes.

From this rule it follows that, in applying the specialization method described above, when we construct the subgraphs $\tilde{G}_k$ and $\tilde{G}_{\bar{k}}$ of $\tilde{G}$ we can omit from $\tilde{G}_{\bar{k}}$ any node x_l such that on $\tilde{G}$, the condition (5.7) holds.

As an illustration, in Fig. 5.6 when we specialized the original problem to obtain $\tilde{G}_1$ and $\tilde{G}_{\bar{1}}$, we could have omitted from $\tilde{G}_{\bar{1}}$ all the nodes x_2, x_3, and x_6; the modified specialization process, in which we have made this simplification, is depicted in Fig. 5.7.

5.2.3. Independent edge sets (matchings)

By analogy with our definition of an independent node set, we say that a set M of edges of a simple graph G is an *independent edge set* or a *matching* of G if no two edges of M are adjacent, that is if no two edges of M have a common endpoint.

As an illustration, Fig. 5.8 shows three different matchings of a simple graph, the edges of the matchings being indicated by bold lines.

Given any matching M on a simple graph $G = (X, E)$ we describe the edges in M as the *matching edges* of G (relative to M), whereas we call the edges in E–M the *non-matching edges* of G (relative to M). The nodes which are endpoints of matching edges are said to be *covered* by M, whereas nodes which are not endpoints of matching edges are said to be *exposed* relative to M.

A matching M is *maximal* if there is no other matching which properly contains M; and M is called a *maximum matching* if no other matching contains more edges. If every node is covered, the matching is said to be *perfect*. Clearly, if a perfect matching exists for a graph G, that matching is a maximum matching.

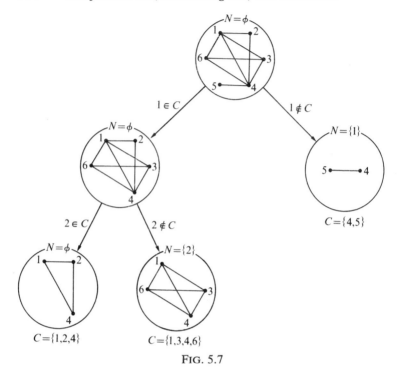

FIG. 5.7

As an illustration, the matchings shown in Fig. 5.8(b) and Fig. 5.8(c) are both perfect. For the graph of Fig. 5.9, the matching shown is a maximum matching; this graph does not have a perfect matching.

Example 5.2. *Matching transistors for push–pull amplifiers.* A batch of power transistors has been manufactured for use in pairs, in push–pull amplifiers. In this application two transistors can only be used together if the differences between their gains, and the differences between their

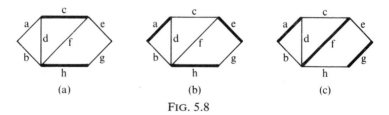

FIG. 5.8

FIG. 5.9

resistances, lie within small prescribed tolerances. Because the fabrication process is imperfect it is necessary to measure the parameters of each of the transistors produced, and on this basis to decide which transistors are compatible. Given this information, how should the transistors be arranged in pairs, to minimize wastage?

Let G be the simple graph whose nodes represent the transistors, two nodes being joined by an edge if the corresponding transistors are compatible. Then a maximum matching on G gives the greatest possible number of 'matched pairs'.

In Section 4.7 we defined the *interchange graph* of a simple graph. From that definition it follows immediately that a set M of edges of a graph G is a matching of G if and only if M is an independent set of nodes on the interchange graph of G. As an illustration, Fig. 5.10 shows the interchange graph for the graph of Fig. 5.8; the independent set marked on Fig. 5.10 corresponds to the matching of Fig. 5.8(b).

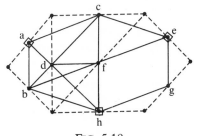

FIG. 5.10

Since the matchings of G correspond to the independent sets of the interchange graph I of G, and these sets in turn correspond to the cliques of the complementary graph $\bar{I}$ of I, it is possible to find all the maximal matchings of G by the algorithm described in the previous section. However if we require only one maximum matching, we can obtain this by an alternative method, of only polynomial complexity, as indicated below.

Alternating chains. Let M be a matching on a graph G. Then a chain κ on G is called an *alternating chain* (relative to M) if (i) the chain κ is simple, and (ii) in each pair of consecutive edges of κ, one edge is a matching edge and the other is a non-matching edge. For instance, in the graph of Fig. 5.8(a) the chain

$$a, c, f, h, g$$

is an alternating chain.

Now let us suppose that, for a given matching M on G, the graph G has an alternating chain κ between two exposed nodes. Let K be the set of edges in κ, and let $\tilde{M}$ be the symmetric difference between M and K,

$$\tilde{M} = M \triangle K,$$

that is, the set of edges which belong either to M or to K but not both. To describe this set in another way, $\tilde{M}$ is the set obtained by deleting from M the matching edges of κ and then adding all the non-matching edges of κ. It is clear from this method of construction that every node which is an endpoint of an edge of M is an endpoint of precisely one edge of $\tilde{M}$; and that none of the nodes which are exposed relative to M are the endpoints of edges of $\tilde{M}$—except each endpoint of κ, which is an endpoint of precisely one edge in $\tilde{M}$. It follows that $\tilde{M}$ is a matching of G, and that this matching has one more edge than M. For this reason, an alternating chain between two exposed nodes is called an *augmenting chain*.

As an illustration, in Fig. 5.8(a) we have depicted a matching

$$M = \{c, h\}.$$

Relative to this matching the graph has two exposed nodes; these are joined by an augmenting chain

$$\kappa = a, c, e$$

from which we obtain the new matching

$$\tilde{M} = \{c, h\} \triangle \{a, c, e\} = \{a, e, h\}$$

which is depicted in Fig. 5.8(b). The graph of Fig. 5.8(a) also has an augmenting chain

$$\kappa = a, c, f, h, g$$

from which we obtain the matching

$$\tilde{M} = \{c, h\} \triangle \{a, c, f, h, g\} = \{a, f, g\}$$

which is depicted in Fig. 5.8(c).

The process of constructing $\tilde{M}$ from M can be visualized as follows. If the edges of M have been drawn as *thick* edges and the edges not in M appear as *thin* edges, then an augmenting chain κ relative to M is a simple chain between two exposed nodes whose edges are alternately thin, thick, thin, ..., thick, thin. To obtain $\tilde{M}$, we simply redraw the edges of κ, replacing its thick edges by thin ones and vice versa.

The above arguments suggest that to obtain a maximum matching, we might first choose a matching M arbitrarily, and then search for an augmenting chain κ relative to M. If this search should be successful we would construct a larger matching $\tilde{M}$, as the symmetric difference between M and the set of edges of κ. Then, after replacing M by $\tilde{M}$, we would search for another augmenting chain, and so on until eventually we obtained a matching M_0 for which no augmenting chains existed. However, can we be sure that, when this algorithm terminates, the final matching M_0 is always a maximum matching? An affirmative answer is provided by the following theorem, due to Berge (1957).

A matching M in a graph G is a maximum matching if and only if G does not contain any augmenting chains relative to M.

To prove the theorem, we need only show that if M is not a maximum matching then G contains an augmenting chain relative to M, the converse having already been established. Let us therefore assume that G has a matching M_0 which contains more edges than M, and let us consider the partial graph $H = (X, F)$ of G, where $F = M \triangle M_0$. Now in H, each node $x_i \in X$ has a node degree $\rho(x_i) \le 2$, since at most one edge of M, and one edge of M_0, is incident with x_i. Thus each connected component of H is either (i) an isolated node or (ii) a circuit of even order, with edges alternately in M and M_0, or (iii) an open simple chain, with edges alternately in M and M_0. Now since M_0 has more edges than M, the graph H has a connected component containing more edges of M_0 than edges of M. It follows that H has a component which is an open simple chain κ, with edges alternately in M and M_0 and whose first and last edges both belong to M_0. On G, the edges of this chain κ form an alternating chain relative to M, and the endpoints of κ are exposed relative to M. Thus G does contain an augmenting chain relative to M, as required.

With regard to the practical determination of augmenting chains, we note first from the proof of Berge's theorem that augmenting chains are always elementary. Now let G be a graph with a matching M, and let $\tilde{G}$ be the graph obtained by adding to G a new node x_0, and placing an edge between x_0 and each node of G which is exposed relative to M (see Fig. 5.11). Also, let us partition the edges of $\tilde{G}$ into two classes—one of *thick* edges (comprising the matching edges of G, together with the new edges incident with x_0), and one of *thin* edges (the non-matching edges of G). Then clearly the augmenting chains of G correspond to those elementary chains from x_0 to itself on $\tilde{G}$, whose edges are alternately thick and thin. (An example of such a chain is indicated by arrows on Fig. 5.11(b).) An elementary chain of this kind can be obtained by a backtrack programming method; alternatively, some very efficient 'node-labelling' algorithms have been devised for this purpose (see the Additional notes and bibliography).

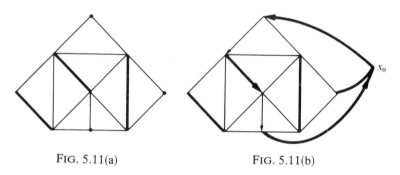

FIG. 5.11(a)   FIG. 5.11(b)

Matching in bipartite graphs. A simple graph $G = (X, E)$ is said to be *bipartite* if its node set X can be partitioned into two subsets X_1 and X_2 such that every edge of G has one endpoint in X_1 and one endpoint in X_2. An example is shown in Fig. 5.12.

Very often, the graphs for which maximum matchings are required arise naturally in a bipartite form—as in the following example.

Example 5.3. *An assignment problem.* A firm has q vacant jobs $\beta_1, \beta_2, \ldots, \beta_q$ of different types. There are p applicants $\alpha_1, \alpha_2, \ldots, \alpha_p$ for work with the firm, each applicant being suited for one or more of the vacancies. How should the applicants be assigned to jobs, in order to fill as many of the jobs as possible?

Let G be the bipartite graph with node set $X = A \cup B$, where A is the set of applicants and B is the set of jobs, and which has an edge $[\alpha_i, \beta_j]$ whenever applicant α_i is able to fill job β_j. Then a matching on G defines an assignment of jobs to applicants; for a maximum matching, the number of jobs assigned is as great as possible. As an illustration, Fig. 5.12 shows a maximum matching on a bipartite graph, of the kind which arises in assignment problems.

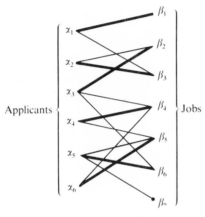

FIG. 5.12

A very efficient algorithm for constructing maximum matchings on bipartite graphs, using augmenting chains, has been devised by Hopcroft and Karp (1973). Problems of this kind can also be solved by the network flow methods which will be presented in the next chapter.

5.3. Dominating sets

Given a simple graph $G = (X, E)$, we say that a set S of nodes of G is a *dominating set* if every node of G either belongs to S or is adjacent to one or more nodes of S. A dominating set is *minimal* if none of its proper subsets are dominating sets.

For the graph of Fig. 5.1, the minimal dominating sets are $\{x_1, x_5\}$, $\{x_2, x_3, x_5\}$, $\{x_2, x_5, x_6\}$, $\{x_4\}$.

The *domination number* $\beta(G)$ of a graph G is the smallest number of nodes in any dominating set of G. For the graph of Fig. 5.1, $\beta(G) = 1$.

As might be expected, there is a close connection between the dominating sets and the independent sets of a graph. In particular, it is easy to prove that *in any graph, an independent set is maximal if and only if it is a dominating set.* Indeed, if I is a maximal independent set then there cannot be any node $x \notin I$ which is not adjacent to some node in I, for otherwise the set $I \cup \{x\}$ would be independent. Conversely, if an independent set I is dominating then it is impossible to add any node to I without destroying the independence of I.

From this result it follows that for any graph, the independence number is greater than or equal to the domination number:

$$\alpha(G) \geq \beta(G).$$

It will be noted that, although a maximal independent set is necessarily a dominating set, a minimal dominating set need not be independent. For instance, the graph of Fig. 5.13 has a minimal dominating set $\{x_3, x_4\}$ which is not independent.

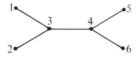

FIG. 5.13

5.4. Colorations

5.4.1. Node colorations

By a *coloration* of a simple graph $G = (X, E)$ we mean a partition $\mathscr{C} = \{X_1, X_2, \ldots, X_k\}$ of its node set X in which every block X_i is an independent set. In pictorial terms, if we suppose that each block of a coloration is associated with a different colour, then we may regard a coloration as an assignment of a colour to each of the nodes of a graph, such that no two adjacent nodes have the same colour.

A coloration which has exactly k blocks is sometimes called a *k-coloration*. The smallest number k for which a graph G has a k-coloration is called the *chromatic number* of G and is denoted by $\gamma(G)$; a coloration which uses only $\gamma(G)$ colours is called a *minimum coloration* of G.

As examples, the chromatic numbers of the graphs of Fig. 5.14 and Fig. 5.15 are 4 and 3 respectively; on these diagrams we have

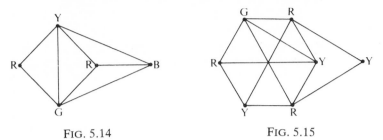

FIG. 5.14   FIG. 5.15

specified minimum colorations by labelling the nodes with the letters R (for *red*), Y (for *yellow*), G (for *green*), and B (for *blue*). As another important example, we note that a complete graph with n nodes has the chromatic number n.

Example 5.4. *Construction of an examination timetable.* In a university faculty, the final-year students have to sit a number of written examinations in different subjects, each examination taking half a day. The examinations in any pair of subjects can be held concurrently if and only if no student is a candidate for both subjects. A timetable is required, in which the examinations are all completed in the shortest possible time.

Let $X = \{x_1, x_2, \ldots, x_n\}$ be the set of all examinations, and let us construct the simple graph $G = (X, E)$, where $[x_i, x_j] \in E$ if and only if the examinations x_i and x_j cannot be held concurrently. Then any coloration of G defines a partition $\mathscr{C} = \{X_1, X_2, \ldots, X_k\}$ of X, such that all the examinations in any one block X_i of $\mathscr{C}$ can be held concurrently. Thus if we find a minimum coloration of G and assign half a day to each block of the corresponding partition of X, we obtain a timetable which meets our requirements.

Example 5.5. *The assignment of memory locations to program variables.* In writing or compiling a program for a computer with a small rapid-access memory it is sometimes desirable to determine which variables may occupy the same memory locations and hence to find an assignment of memory locations to variables which uses least space. To achieve this we may construct a *data-transmission graph* G of a program as follows:

(i) Each instruction of the program is represented in G by a pair of nodes r_k and t_k, joined by an arc (r_k, t_k). The nodes r_k and t_k are called respectively the *receiver* and *transmitter* nodes of the kth program instruction. The node r_k is labelled with the set of program variables whose values are read (or *received*) from the memory in the execution of the kth instruction; whereas the node t_k is labelled with the set of variables to which values are assigned (and *transmitted* to the memory) in executing the instruction.

(ii) If the *i*th instruction transfers control (conditionally or unconditionally) to the *j*th instruction, then G has an arc (t_i, r_j).

As a simple example, Fig. 5.16(a) is the flow chart of a program to calculate the highest common factor of two integers m and n, by Euclid's method. (In this program, first the number m is divided by n, to obtain their quotient $q = \lfloor m/n \rfloor$ and corresponding remainder r. If the remainder is non-zero, the value of n is assigned to m, the value of r is assigned to n, and the division process is repeated. Eventually a zero remainder is obtained, at which point the value of n is the highest common factor of the original pair of numbers.) The corresponding data-transmission graph is shown in Fig. 5.16(b).

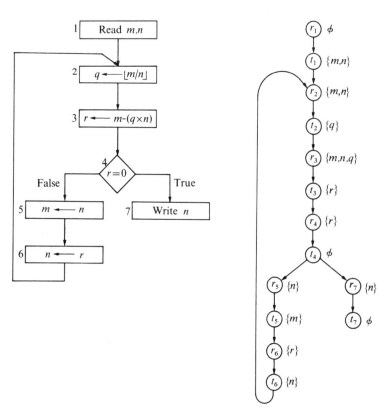

FIG. 5.16(a)   FIG. 5.16(b)

On a data-transmission graph G we define a *carrier of a variable v* as a path from a transmitter of v to a receiver of v, on which none of the intermediate nodes are transmitters of v. Now let G_v be the graph comprising the nodes and arcs of G which belong to carriers of v; we describe the node-set of each connected component of G_v as a *region of v*.

As an illustration, the program by Fig. 5.16(a) has four variables, m, n, q, and r. The corresponding graphs G_m, G_n, G_q and G_r are shown in Fig. 5.16(c); since these graphs are all connected, our program has only one region for each variable.

Now let S be the set of regions of all the variables of a program and let M be the set of memory locations of a computer. Then a *memory assignment* is a function $f: S \to M$ which assigns a memory location to each region. A memory assignment f is said to be *proper* if for any two regions $s_i, s_j \in S$ such that $s_i \cap s_j \neq \phi$, $f(s_i) \neq f(s_j)$. Thus with any proper memory assignment, the contents of a memory location can only be 'over-written' when they are no longer required.

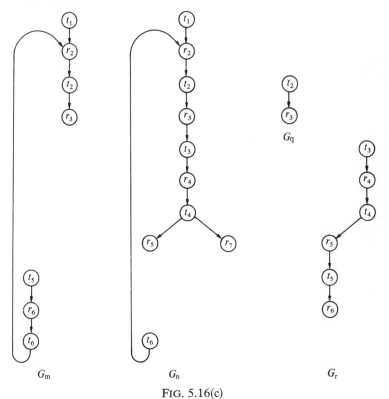

FIG. 5.16(c)

To obtain a proper memory assignment we may construct an *interference graph* H; this is a simple graph, whose node set is the set S of program regions, two nodes s_i and s_j being joined by an edge whenever $s_i \cap s_j \neq \phi$. Then the colorations of H correspond to the proper memory assignments for the program, and in particular a minimum coloration of H determines a proper memory assignment using least memory locations.

For the program of Fig. 5.16(a), the interference graph is shown in Fig. 5.16(d). (Since this particular program has only one region for each variable, we have simply labelled the nodes (regions) in H with the names of the corresponding variables.) The graph has a 3-coloration, hence only 3 memory locations are required—one to store m, one to store n, and one for the pair of variables q and r. (Alternatively, we might 'rename' q as r, or r as q, throughout the program text.)

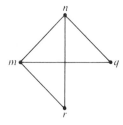

FIG. 5.16(d)

5.4.2. An algorithm for colouring a graph

The colorations of a graph G can be constructed by backtrack programming, as follows.

If the graph G is complete then it has only one coloration (in which every node is assigned a different colour). Otherwise, the problem of finding all the colorations can be 'specialized' (cf. Section 2.7) in the following way. Let x_i and x_j be any two nodes of G which are not adjacent; then the set S of colorations of G can be partitioned into two sets S_{ij} and $S_{\overline{ij}}$, where S_{ij} is the set of all colorations in which x_i and x_j are of the same colour, and $S_{\overline{ij}}$ is the set of all colorations in which x_i and x_j have different colours. Clearly, the colorations in S_{ij} correspond to the colorations of the condensation G_{ij} of G which is obtained by coalescing x_i and x_j; whereas $S_{\overline{ij}}$ is the set of colorations of the graph $G_{\overline{ij}}$ obtained by adding the edge $[x_i, x_j]$ to G.

It is evident that the graph G_{ij} has one node less than G, while $G_{\overline{ij}}$ has the same node set as G, but one more edge. Thus, by repetition of the specialization process we ultimately obtain graphs which are all complete, and the (unique) colorations of all these complete graphs give all the colorations of the original graph *G*.

This specialization process is demonstrated for a 5-node graph in Fig. 5.17. (In this diagram, the symbolism $x_i \equiv x_j$ indicates that x_i has the same colour as x_j.) It will be seen that the graph has a chromatic number of 3; it has one 3-coloration, namely

$$\{\{x_1, x_3\}, \{x_2, x_4\}, \{x_5\}\},$$

and also three 4-colorations, and one 5-coloration.

It is often possible to simplify graphs obtained during the search, by applying the following rule.

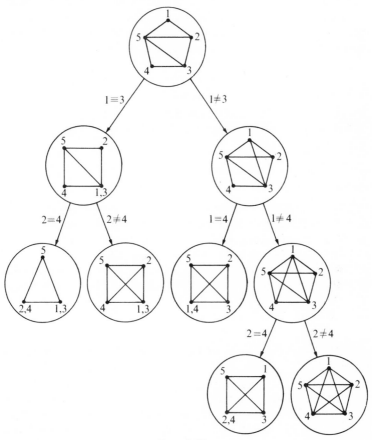

FIG. 5.17

Rule 1: If a graph G has a node x_i which is adjacent to every other node, then in every coloration of G the node x_i has a colour different from the colours of all other nodes.

Thus, a node which is adjacent to all other nodes can be deleted, for the purpose of finding colorations. Simplifications of this kind do not reduce the number of specializations to be performed, but they greatly reduce the amount of data to be manipulated and stored at each stage. For the problem of Fig. 5.17 for instance, *Rule 1* enables us to remove node x_5 from the original problem, and to make further simplifications to all the sub-problems.

Determination of a minimum coloration. In practical problems we usually require only one coloration, which uses least colours. To find such a coloration, it is unfortunately still necessary to use a search method of the kind described above, and to extract a minimum coloration from the set of all colorations obtained. However, it is usually possible to 'prune' the search tree in such a way as to discard many sub-problems, while retaining always at least one sub-problem which leads to a minimum coloration.

For instance, we can make use of the following rule.

Rule 2: If a graph G has two nodes x_i and x_j such that $\Gamma(x_i) \subseteq \Gamma(x_j)$, then G has a minimum coloration in which x_i and x_j are of the same colour.

Thus, if a graph G has two nodes x_i and x_j with $\Gamma(x_i) \subseteq \Gamma(x_j)$, the problem of finding a minimum coloration of G reduces to the problem of finding a minimum coloration of its condensation G_{ij} (which is obtained simply by removing x_i from G).

As an example, it follows from *Rule 2* that the 5-node graph of Fig. 5.17 has a minimum coloration in which x_1 has the same colour as x_3, and x_2 has the same colour as x_4; by constructing the corresponding condensation, we immediately obtain the complete 3-node graph which defines its minimum coloration. Minimum colorations of the graphs of Fig. 5.14 and Fig. 5.15 can also be obtained easily, by repeated applications of *Rule 2.*

As a further method of pruning the search tree, one can use a 'branch-and-bound' method, in which the search is restricted by calculating lower bounds to the chromatic numbers of graphs obtained in the course of the specialization process (Corneil and Graham 1973).

Finally, we note that if at any stage we obtain a graph which has articulation nodes, a minimum coloration of that graph can be obtained by finding a minimum coloration of each of its blocks separately.

5.4.3. Edge colorations

An *edge coloration* of a simple graph $G = (X, E)$ is a partition $\mathscr{C} = \{E_1, E_2, \ldots, E_k\}$ of its edge set E in which every block E_i is a matching of G. In pictorial terms, if we suppose that each block of an edge coloration is associated with a particular colour, then we may consider an edge coloration as an assignment of a colour to every edge of a graph, such that no two adjacent edges have the same colour.

An edge coloration which has exactly k blocks is sometimes called a *k-edge coloration*. The smallest number k for which a graph G has a k-edge coloration is called the *chromatic index* $\chi(G)$ of G.

From these definitions it follows immediately that every k-edge coloration of a graph G determines a k-node coloration of its interchange graph, and that the converse is also true. (As an illustration, Fig. 5.18(a) shows a minimum edge coloration of a graph, and Fig. 5.18(b) shows the corresponding minimum node coloration of its interchange graph.) It is therefore possible to construct edge colorations by the algorithms described in the previous section.

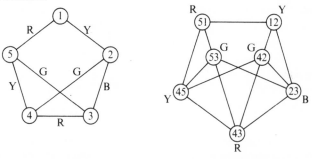

FIG. 5.18(a)   FIG. 5.18(b)

Example 5.6. *A wiring problem.* An electronic unit consists of a number of integrated circuit modules, joined together by coloured wires. To facilitate testing of the unit, it is required that all wires connected to the same module be coloured differently. How can this be achieved, using as few different colours of wire as possible?

In graph-theoretic terms, this problem simply involves finding a minimum edge coloration of the graph whose nodes represent the modules and whose edges represent the wires joining them.

Exercises

5.1. A firm manufactures Schottky diodes, to be used in 'matched pairs' in radar sets. For a batch of ten diodes, measurements have been made of certain diode parameters (in particular, their junction capacitances and series resistances), and from these measurements it has been found that the following pairs of diodes are compatible:

$$[1, 2], [1, 5], [2, 3], [2, 4], [2, 5], [3, \ 5], [4, 5], [4, 6], [4, \ 9],$$
$$[5, 6], [5, 7], [6, 7], [6, 8], [6, 9], [6, 10], [7, 9], [8, 9], [9, 10].$$

How many matched pairs can be obtained from the batch?

5.2. Prove that, if the largest node degree of a simple graph is ρ, then its chromatic number is not greater than $\rho + 1$.

5.3. Prove that a graph has a 2-coloration if and only if it does not contain any circuits of odd length.

5.4. Seven local television stations are to be built in different geographic locations. Stations which are far apart can use the same frequency, but to avoid interference certain pairs of stations must use different frequencies; these pairs are specified by crosses in the table below.

	1	2	3	4	5	6	7
1		×	×		×		×
2	×		×				
3	×	×		×	×		×
4			×		×	×	
5	×		×	×		×	×
6				×	×		
7	×		×		×		

Find an assignment of frequencies to transmitters which uses the smallest possible number of different frequencies.

Additional notes and bibliography

It has been shown by Moon and Moser (1965) that the number of cliques in a graph may grow exponentially with n. However in many practical applications this does not happen, and the specialization method of Section

5.2.2 for constructing the maximal cliques works very effectively (see Bron and Kerbosch 1973). Some other algorithms for constructing maximal cliques are described by Augustson and Minker (1970), Mulligan and Corneil (1972), Akkoyunlu (1973), Osteen and Tou (1973), Osteen (1974), and Johnston (1976).

The first polynomial algorithm for constructing a maximum matching was developed by Edmonds (1965); this algorithm was of complexity $O(n^4)$. More efficient 'node labelling' techniques for finding augmenting chains were subsequently developed by Witzgall and Zahn (1965), Balinski (1969), Even and Kariv (1975), and Gabow (1976). The algorithm of Even and Kariv is of least complexity, this being only $O(n^{2.5})$.

For a full discussion of the analytical results obtained relating to node and edge colorations, see Berge (1976) and Ore (1967).

The method of assigning memory locations to variables described in Example 5.7 is based on the work of Lavrov (1961). See also Logrippo (1972, 1978).

The backtrack programming method of constructing colorations (Section 5.4.1) is based on a technique of Zykov (1949) for representing chromatic polynomials; the simplification rules were suggested by Hedetniemi (1971). An alternative coloration method involving the generation of maximal independent sets was developed by Christofides (1971) and subsequently refined by Roschke and Furtado (1973) and Wang (1974). Corneil and Graham (1973) compared their branch-and-bound method (based on Zykov's technique) with the Roschke–Furtado algorithm for a number of families of graphs, and found that the branch-and-bound method was always substantially superior.

The problem of finding a minimum coloration is NP-complete (see Karp 1972; Lawler 1976b). Polynomial algorithms have been developed, to construct 'approximate' solutions to the minimum coloration problem (see Peck and Williams 1966; Welsh and Powell 1967; Wood 1969; Williams 1970; Matula, Marble, and Isaacson 1972). However, it has subsequently been shown that the results of all these algorithms may be arbitrarily bad (Johnson 1974; Mitchem 1976; Garey and Johnson 1976).

In practice it is very difficult to obtain sharp bounds for the chromatic number of a graph, but very sharp bounds for the chromatic index are easily obtainable: specifically, if ρ is the largest node degree of a simple graph G, then $\rho \leq \chi(G) \leq \rho + 1$. The validity of the first of these inequalities is evident; the proof of the second (which is due to Vizing) is given in English by Ore (1967) and Berge (1976).

It is possible to enumerate the maximal independent sets and cliques, the minimal dominating sets, and the colorations of a graph by Boolean methods (Hammer and Rudeanu 1968). See also Kaufmann and Pichat (1977).

6 Flows in networks

6.1. Introduction

THIS CHAPTER is concerned with p-graphs in which some substance can flow along the arcs, from one point to another. Much of the theory presented here originated in the study of transportation problems, i.e. problems of transporting a commodity from certain points of supply to other points of demand in such a way as to minimize shipping cost. However, the graph-theoretic techniques first developed in this context are applicable to several other kinds of flow problems, involving for instance the flow of information in communication systems, and the flow of traffic in road networks. Furthermore, many practical problems of a combinatorial nature, which do not involve flows in any physical sense, can nevertheless be formulated and solved very elegantly by using network-flow models.

6.2. Networks

In discussing flows it will be convenient to consider only p-graphs which are connected and which do not contain any loops. This will not involve any loss in generality, for the flows in a disconnected p-graph can be analysed by considering each of its connected components separately, and flows in loops contribute nothing to flows between nodes. In this chapter, p-graphs which are connected and without loops will be called *networks*.

Now let $G = (X, U)$ be a network, whose nodes and arcs are arbitrarily numbered: $X = \{x_1, x_2, \ldots, x_n\}$ and $U = \{u_1, u_2, \ldots, u_m\}$. Then the *incidence matrix* of G is the $n \times m$ matrix $S = [s_{ij}]$ whose rows and columns correspond to the nodes and arcs of G respectively, and whose elements are

$$s_{ij} = \begin{cases} +1 & \text{if } u_j \text{ is incident from } x_i, \\ -1 & \text{if } u_j \text{ is incident to } x_i, \\ 0 & \text{if } u_j \text{ is not incident to or from } x_i. \end{cases}$$

As an example, the network of Fig. 6.1 has the incidence matrix

$$S = \begin{bmatrix} +1 & +1 & 0 & -1 & 0 & 0 & 0 & -1 & 0 \\ -1 & 0 & +1 & 0 & -1 & 0 & 0 & 0 & 0 \\ 0 & 0 & 0 & +1 & +1 & -1 & -1 & 0 & 0 \\ 0 & -1 & 0 & 0 & 0 & +1 & +1 & 0 & -1 \\ 0 & 0 & -1 & 0 & 0 & 0 & 0 & +1 & +1 \end{bmatrix}.$$

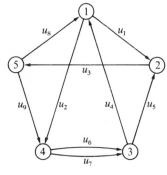

FIG. 6.1   FIG. 6.2

6.3. Network flows

6.3.1. Definition of a network flow

A *flow* on a network G is a vector $\mathbf{f} = [f_1, f_2, \ldots, f_m]$ of m real numbers (where m is the number of arcs in G) such that

(i) each element f_i of $\mathbf{f}$, which is called the *flow in the arc* u_i, is non-negative, and

(ii) for every node x_i of G, the sum of the flows in arcs incident to x_i is equal to the sum of the flows in arcs incident from x_i. (This condition is called the *flow conservation condition*.)

The condition (i) is conveniently written as

$$\mathbf{f} \geq 0; \tag{6.1}$$

the flow conservation condition (ii) can be expressed in terms of the coefficients of the incidence matrix of G, as

$$\sum_{j=1}^{m} s_{ij} f_j = 0, \qquad (i = 1, 2, \ldots, n), \tag{6.2(a)}$$

or, more concisely, as

$$Sf' = 0. \qquad (6.2(b))$$

As an example, the vector $[1, 1, 3, 1, 2, 2, 1, 1, 2]$ is a flow on the network of Fig. 6.1; this flow is depicted in Fig. 6.2.

6.3.2. Operations on flows

Let f be a flow on a network G, and let k be a non-negative number. Then kf is a flow on G, since the condition $f \geq 0$ implies that $kf \geq 0$, and the condition $Sf' = 0$ implies that $S(kf)' = Sf'k = 0$.

In the same way, it is easily verified that for any two flows f_1 and f_2 on the same network G, the vector sum $f_1 + f_2$ is a flow on G; and if $f_1 \geq f_2$ then the difference $f_1 - f_2$ is also a flow on G.

6.3.3. Elementary flows

Let γ be an elementary cycle on G, and let v be the vector with elements

$$v_i = \begin{cases} 1 & \text{if arc } u_i \text{ lies on } \gamma, \\ 0 & \text{otherwise,} \end{cases} \qquad (i = 1, 2, \ldots, m).$$

It is easily verified that v is a flow on G: this flow is called the *elementary cyclic flow* associated with γ.

Now let $v_1, v_2, \ldots, v_k$ be elementary cyclic flows on G and let $r_1, r_2, \ldots, r_k$ be non-negative numbers. Then, from our previous discussion of operations on flows it is clear that the vector

$$f = r_1 v_1 + r_2 v_2 + \cdots + r_k v_k \qquad (6.3)$$

is a flow on G.

Conversely, any feasible flow f on G can be expressed in the form (6.3). To prove this, let us suppose that f is a non-zero flow on G (for a zero flow, the proof is trivial), and let H be the partial graph of G obtained by deleting those arcs in which the flow is zero. Now since $f \neq 0$, H contains at least one arc; and from the flow conservation condition, it follows that on H

$$\rho^+(x_i) = 0 \quad \text{if and only if} \quad \rho^-(x_i) = 0 \quad \text{for all } x_i \in X.$$

Hence, starting from the initial endpoint of any arc of H, it is possible to construct a path of arbitrary order on H, which implies that H contains at least one elementary cycle. Now let γ_1 be any elementary cycle on H, let r_1 be the smallest flow in the arcs of γ_1, and let $\mathbf{v}_1$ be the elementary cyclic flow on G associated with γ_1. Then $\mathbf{f}_1 = \mathbf{f} - r_1\mathbf{v}_1$ is a flow on G, having more zero elements than $\mathbf{f}$. By repeated decompositions of this kind we ultimately obtain a flow

$$\mathbf{f}_k = \mathbf{f} - r_1\mathbf{v}_1 - r_2\mathbf{v}_2 \cdots - r_k\mathbf{v}_k = 0,$$

from which it follows that $\mathbf{f}$ is expressible in the form (6.3).

The fact that any network flow can be decomposed into elementary cyclic flows—or constructed by combining elementary flows—will be of great importance in later sections.

As an example, Fig. 6.3 depicts two different decompositions of the network flow of Fig. 6.2.

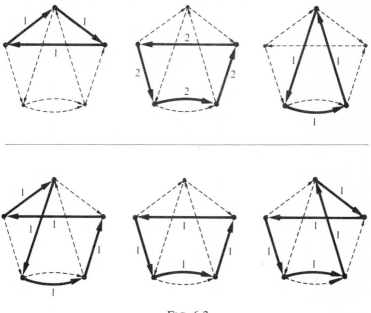

FIG. 6.3

6.3.4. Arc capacities

Let us now suppose that each arc u_i of G has associated with it a non-negative integer c_i, called the *capacity* of u_i; this may be regarded as the maximum permissible value of the flow in the arc u_i. Then a flow $\mathbf{f}$ on G is said to be *feasible* if and only if

$$f_i \le c_i, \qquad (i = 1, 2, \ldots, m). \tag{6.4}$$

To express the feasibility condition in matrix form, we define the *capacity vector* of G as $\mathbf{c} = [c_1, c_2, \ldots, c_m]$. Then from (6.4) a flow $\mathbf{f}$ on G is feasible if and only if

$$\mathbf{f} \le \mathbf{c}. \tag{6.5}$$

The properties of feasible flows are most conveniently described in terms of 'displacement networks', which are defined in the next section.

6.4. Displacement networks

6.4.1. *The notion of a displacement network*

Let $\mathbf{f}$ be a feasible flow on G; then the *displacement network* $\tilde{G}(\mathbf{f})$ associated with $\mathbf{f}$ is the network which has the same nodes as G, and arcs determined as follows. For each arc u_i of G, $\tilde{G}(\mathbf{f})$ has (i) a *normal arc* u_i^+ which has the same initial and terminal endpoints as u_i, and (ii) an *inverted arc* u_i^- which has the same endpoints as u_i but the opposite orientation. The capacities of u_i^+ and u_i^-, which are denoted by c_i^+ and c_i^- respectively, are defined by

$$\left. \begin{array}{l} c_i^+ = c_i - f_i, \\ c_i^- = f_i, \end{array} \right\} \quad (i = 1, 2, \ldots, m). \tag{6.6}$$

For example, Fig. 6.5 shows the displacement network associated with the network flow of Fig. 6.4; in Fig. 6.5 the solid lines represent normal arcs, broken lines represent inverted arcs.

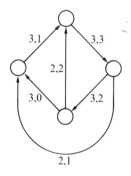

FIG. 6.4.The first number on each arc is its capacity; the second is the arc flow.

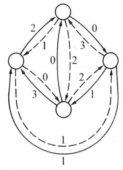

FIG. 6.5. The numbers indicate arc capacities.

It will be observed that on a displacement network $\tilde{G}(\mathbf{f})$, the capacity of each normal (or, respectively, inverted) arc is the amount by which the flow in the corresponding arc of the original network G can be increased (or, respectively, reduced) without exceeding the arc capacity (or becoming negative). As one might therefore expect, for a given flow $\mathbf{f}$ on G it is possible to describe the 'difference' between $\mathbf{f}$ and any other flow on G in terms of a feasible flow on the displacement network $\tilde{G}(\mathbf{f})$. To demonstrate how this can be done it will be helpful first to establish some properties of the feasible flows on $\tilde{G}(\mathbf{f})$.

6.4.2. *Flows on displacement networks*

Let us suppose that the arcs of $\tilde{G}(\mathbf{f})$ are listed in the order $u_1^+, u_2^+, \ldots, u_m^+, u_1^-, u_2^-, \ldots, u_m^-$. Then the incidence matrix $\tilde{S}$ of $\tilde{G}(\mathbf{f})$, which is of dimensions $n \times 2m$, can be written in the partitioned form

$$\tilde{S} = [S \,|\, -S], \tag{6.7}$$

where S is the $n \times m$ incidence matrix of G. The capacity vector $\tilde{\mathbf{c}}$ of $\tilde{G}(\mathbf{f})$ is also conveniently written in the partitioned form

$$\tilde{\mathbf{c}} = [\mathbf{c}^+ \,|\, \mathbf{c}^-], \tag{6.8}$$

where $\mathbf{c}^+ = [c_1^+, c_2^+, \ldots, c_m^+]$ and $\mathbf{c}^- = [c_1^-, c_2^-, \ldots, c_m^-]$. From (6.6) these two components of $\tilde{\mathbf{c}}$ can be written as

$$\left. \begin{aligned} \mathbf{c}^+ &= \mathbf{c} - \mathbf{f}, \\ \mathbf{c}^- &= \mathbf{f}. \end{aligned} \right\} \tag{6.9}$$

Now from the definition of a feasible flow it follows that a vector $\tilde{\mathbf{f}}$ of $2m$ real numbers is a feasible flow on $\tilde{G}(\mathbf{f})$ if and only if

$$0 \leq \tilde{\mathbf{f}} \leq \tilde{\mathbf{c}} \qquad 6.10)$$

and

$$\tilde{S}\tilde{\mathbf{f}}' = 0. \qquad (6.11)$$

If $\tilde{\mathbf{f}}$ is partitioned in the form

$$\tilde{\mathbf{f}} = [\mathbf{f}^+ | \mathbf{f}^-], \qquad (6.12)$$

where $\mathbf{f}^+$ and $\mathbf{f}^-$ each have m elements, then the condition (6.10) can be written as

$$\left.\begin{array}{l} 0 \leq \mathbf{f}^+ \leq \mathbf{c}^+, \\ 0 \leq \mathbf{f}^- \leq \mathbf{c}, \end{array}\right\} \qquad (6.13)$$

and by (6.9) these conditions can be expressed as

$$\left.\begin{array}{l} 0 \leq \mathbf{f}^+ \leq \mathbf{c} - \mathbf{f}, \\ 0 \leq \mathbf{f}^- \leq \mathbf{f}. \end{array}\right\} \qquad (6.14)$$

Also, by (6.7) and (5.12), the conservation condition (6.11) can be expressed in the form

$$S(\mathbf{f}^+ - \mathbf{f}^-)' = 0. \qquad (6.15)$$

Thus, a vector $\tilde{\mathbf{f}} = [\mathbf{f}^+ | \mathbf{f}^-]$ is a feasible flow on $\tilde{G}(\mathbf{f})$ if and only if its components $\mathbf{f}^+$ and $\mathbf{f}^-$ satisfy the conditions (6.14) and (6.15).

6.4.3. *Flow differences*

Now let $\mathbf{f}$ and $\mathbf{g}$ be any two feasible flows on G. Then the vector difference $\mathbf{g} - \mathbf{f}$ determines a feasible flow on $\tilde{G}(\mathbf{f})$, as follows: Let $\mathbf{f}^+$ and $\mathbf{f}^-$ be the vectors with elements

$$f_i^+ = \begin{cases} g_i - f_i & \text{if } g_i \geq f_i, \\ 0 & \text{otherwise,} \end{cases} \quad \text{and} \quad f_i^- = \begin{cases} 0 & \text{if } g_i \geq f_i, \\ f_i - g_i & \text{otherwise,} \end{cases}$$

$$(i = 1, 2, \ldots, m). \qquad (6.16)$$

It is evident that

$$\mathbf{f}^+ - \mathbf{f}^- = \mathbf{g} - \mathbf{f}. \qquad (6.17)$$

Also, it is easily verified that the vector $\tilde{\mathbf{f}} = [\mathbf{f}^+ | \mathbf{f}^-]$ is a feasible flow on $\tilde{G}(\mathbf{f})$. Indeed, since $0 \leq \mathbf{f} \leq \mathbf{c}$ and $0 \leq \mathbf{g} \leq \mathbf{c}$ it follows immediately from (6.16) that the vectors $\mathbf{f}^+$ and $\mathbf{f}^-$ satisfy the feasibility conditions (6.14); and by (6.17),

$$S(\mathbf{f}^+ - \mathbf{f}^-)' = S(\mathbf{g} - \mathbf{f})' = S\mathbf{g}' - S\mathbf{f}' = 0$$

so the flow conservation condition (6.15) is also satisfied.

It will be noted that the flow $\tilde{\mathbf{f}} = [\mathbf{f}^+ | \mathbf{f}^-]$ on $\tilde{G}(\mathbf{f})$ defined by (6.16) has the property

$$f_i^+ \cdot f_i^- = 0, \qquad (i = 1, 2, \ldots, m);$$

a flow with this property is said to be *disjunctive*.

As an illustration, Figs. 6.4 and 6.6 show two flows $\mathbf{f}$ and $\mathbf{g}$ on the same network; the disjunctive flow on $\tilde{G}(\mathbf{f})$ associated with the difference $\mathbf{g} - \mathbf{f}$ is shown in Fig. 6.7.

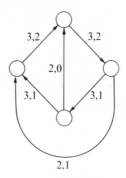

FIG. 6.6. The first number on each arc is its capacity; the second is the arc flow.

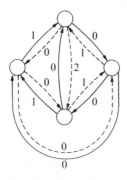

FIG. 6.7. The numbers indicate arc flows.

It has been demonstrated that if **f** and **g** are feasible flows on G, then the vector $\tilde{\mathbf{f}} = [\mathbf{f}^+ | \mathbf{f}^-]$ defined by (6.16) is a disjunctive feasible flow on $\tilde{G}(\mathbf{f})$. Conversely, it can be shown that if **f** is a feasible flow on G, and $\tilde{\mathbf{f}} = [\mathbf{f}^+ | \mathbf{f}^-]$ is a feasible flow on $\tilde{G}(\mathbf{f})$, disjunctive or otherwise, then the vector **g** defined by (cf. (6.17))

$$\mathbf{g} = \mathbf{f} + \mathbf{f}^+ - \mathbf{f}^- \qquad (6.18)$$

is a feasible flow on G. Indeed, it follows from (6.14) that

$$-\mathbf{f} \leq \mathbf{f}^+ - \mathbf{f}^- \leq \mathbf{c} - \mathbf{f} \qquad (6.19)$$

and by adding **f** throughout (6.19) we obtain the feasibility condition

$$0 \leq \mathbf{g} \leq \mathbf{c}.$$

Also, using (6.15) we obtain the flow conservation condition

$$S\mathbf{g}' = S(\mathbf{f} + \mathbf{f}^+ - \mathbf{f}^-)' = S\mathbf{f}' + S(\mathbf{f}^+ - \mathbf{f}^-)' = 0$$

as required.

6.5. Maximal flows in networks

6.5.1. The maximal flow problem

It will now be supposed that G contains two nodes x_s and x_t, called the *source* and *sink* of G respectively, which are joined by an arc $u_r = (x_t, x_s)$ of infinite capacity; this arc is called the *return arc* of G. For any feasible flow **f** on G, we describe the flow f_r in the return arc u_r as the *value* of **f**. The problem to be considered here is that of finding a *maximal flow*, that is, a feasible flow whose value is as large as possible.

Example 6.1. *A shipping problem.* A certain commodity is stored at p depots $\alpha_1, \alpha_2, \ldots, \alpha_p$, each depot α_i having a stock of s_i units. The commodity is required at q distribution centres $\beta_1, \beta_2, \ldots, \beta_q$, the demand at β_j being for d_j units. The maximum quantity w_{ij} which can be transported from each depot α_i to each distribution centre β_j is specified. Is it possible to meet all the demands? How many units of the commodity should be sent from each depot to each distribution centre, in order to meet the demands?

This problem can be represented by a network of the form shown in Fig. 6.8, in which the labels on arcs indicate their capacities. For any maximal flow on this network, the arc flows define a distribution of the commodity

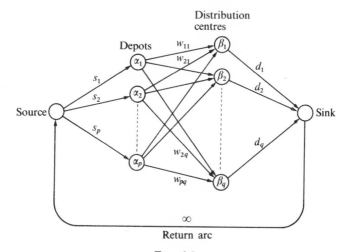

FIG. 6.8

from the depots α_i to the centres β_j which meets the demands as well as possible.

It was shown earlier that any flow on a network can be expressed as a combination of elementary cyclic flows. This suggests that to obtain a maximal flow we might successively combine elementary flows, in such a way as to increase the value of the resulting flow at every stage, until we obtain a flow whose value is maximal. But how should these elementary flows be chosen?

Let $\mathbf{f}$ be any flow on a network G; then we define a *flow-augmenting cycle* of the corresponding displacement network $\tilde{G}(\mathbf{f})$ as an elementary cycle which traverses u_r^+ but not u_r^-, and whose arcs all have non-zero capacities. The significance of flow-augmenting cycles is established in the following theorem:

A flow $\mathbf{f}$ on a network G is a maximal flow if and only if $\tilde{G}(\mathbf{f})$ does not contain any flow-augmenting cycles.

To prove the theorem, let us first suppose that $\tilde{G}(\mathbf{f})$ contains a flow-augmenting cycle γ, and let δ be the capacity of γ (i.e. the smallest of its arc capacities). Also, let $\tilde{\mathbf{f}} = [\mathbf{f}^+ | \mathbf{f}^-]$ be the feasible flow on $\tilde{G}(\mathbf{f})$ obtained by assigning a flow of δ units to each arc of γ. Since u_r^+ lies on γ and u_r^- does not lie on γ,

$$\mathbf{f}_r^+ = \delta \quad \text{and} \quad \mathbf{f}_r^- = 0. \tag{6.20}$$

Now let **g** be the feasible flow on G defined by (cf. 6.18)

$$\mathbf{g} = \mathbf{f} + \mathbf{f}^+ - \mathbf{f}^-. \tag{6.21}$$

From (6.20) and (6.21) it follows that

$$g_r = f_r + f_r^+ - f_r^- = f_r + \delta. \tag{6.22}$$

Thus the value of the flow **g** is δ units greater than the value of **f**, which implies that **f** is not a maximum flow.

Conversely, if **f** is not a maximum flow, there exists a feasible flow **g** of G whose value is greater than that of **f**:

$$g_r > f_r. \tag{6.23}$$

Now let $\tilde{\mathbf{f}} = [\mathbf{f}^+ | \mathbf{f}^-]$ be the flow on $\tilde{G}(\mathbf{f})$ determined by the flow difference $\mathbf{g} - \mathbf{f}$, through the rule (6.16), and let us express $\tilde{\mathbf{f}}$ in the form

$$\tilde{\mathbf{f}} = r_1 \tilde{\mathbf{v}}_1 + r_2 \tilde{\mathbf{v}}_2 + \cdots + r_k \tilde{\mathbf{v}}_k \tag{6.24}$$

where $r_1, r_2, \ldots, r_k$ are positive numbers and $\mathbf{v}_1, \tilde{\mathbf{v}}_2, \ldots, \tilde{\mathbf{v}}_k$ are elementary cyclic flows on $\tilde{G}(\mathbf{f})$. Since $\tilde{\mathbf{f}}$ is a disjunctive flow, it follows from (6.23) that

$$f_r^+ > 0 \quad \text{and} \quad f_r^- = 0.$$

Hence in the flow decomposition (6.24) there exists at least one elementary flow $\tilde{\mathbf{v}}_i$ say for which the flow in the arc u_r^+ is non-zero and the flow in u_r^- is zero. The corresponding elementary cycle γ_i on $\tilde{G}(\mathbf{f})$ therefore traverses u_r^+ but not u_r^-. Also, since $\tilde{\mathbf{f}}$ is a feasible flow on $\tilde{G}(\mathbf{f})$, the flow $\tilde{\mathbf{v}}_i$ is a feasible flow on $\tilde{G}(\mathbf{f})$, which implies that every arc on γ_i has a non-zero capacity. It follows that γ_i is a flow-augmenting cycle on $\tilde{G}(\mathbf{f})$, which proves the theorem.

Example 6.2. Let us consider the network flow **f** which is depicted in Fig. 6.9. (In this diagram, the first number on each arc is its capacity, the second is the arc flow; the bold lines indicate arcs which are *saturated*, i.e. arcs in which the flow is equal to the capacity.) At first sight this flow—which is of value 6—might appear to be a maximal flow, but we shall see that this is not the case. In Fig. 6.10 we have drawn the corresponding displacement network $\tilde{G}(\mathbf{f})$. (In this diagram the arc labels represent their capacities; for simplicity, the arcs of zero capacity have been omitted.) It will be observed that $\tilde{G}(\mathbf{f})$ has a flow-augmenting cycle

$$\gamma = (x_1, x_2), (x_2, x_3), (x_3, x_4), (x_4, x_6), (x_6, x_1),$$

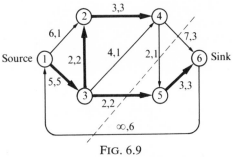

FIG. 6.9

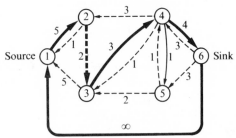

FIG. 6.10

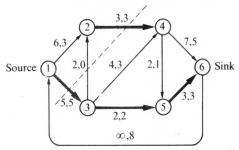

FIG. 6.11

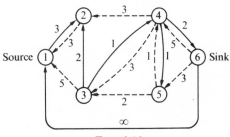

FIG. 6.12

of capacity 2. (On Fig. 6.10 the arcs of this cycle are drawn in bold lines.) To obtain the augmented flow **g**, as defined by (6.21), we modify the flow **f** as follows: for each *normal* arc u_j^+ on γ we *increase* the flow in the corresponding arc u_j on G by 2 units, and for each *inverted* arc u_j^- of γ we *decrease* the flow in u_j on G by 2 units. The resulting flow, of value 8, is shown in Fig. 6.11.

The displacement network associated with this new flow is shown in Fig. 6.12. Since this does not contain any flow-augmenting cycles, the flow depicted in Fig. 6.11 is a maximal flow.

To summarize our discussion so far, we have shown that for any flow **f** on a network G, it is possible to determine from $\tilde{G}(\mathbf{f})$ whether or not the flow is maximal. It has also been demonstrated that if **f** is not a maximal flow, we can construct a flow of larger value. However, it is not yet clear that the repetition of our flow-augmentation method will always yield a maximal flow in a finite number of steps. To demonstrate this we must first examine how the maximum value of the flows in a network is determined, by the capacities of its arcs.

6.5.2. Cuts

The notion of a cut of a graph, which was presented in Section 4.2.2, can be extended to networks in an obvious way: Let $G = (X, U)$ be a network, and let $\{X', X''\}$ be any partition of its node-set X; then the set of all arcs of G whose initial endpoints belong to X' and whose terminal endpoints belong to X'' is called a *cut* of G, and is denoted by (X', X''). For any two nodes $x_i, x_j \in X$, a cut (X', X'') such that $x_i \in X'$ and $x_j \in X''$ is said to *separate* x_i *from* x_j (in that order).

On a network, we define the *capacity* of a cut (X', X'') as the sum of the capacities of its arcs,

$$\sum_{\{j|u_j\in(X',X'')\}} c_j.$$

As an illustration, the broken line on the network of Fig. 6.9 indicates a cut separating its source from its sink. Here $X' = \{x_1, x_2, x_3, x_4\}$, $X'' = \{x_5, x_6\}$ and

$$(X', X'') = \{(x_3, x_5), (x_4, x_5), (x_4, x_6)\}.$$

(It will be noted that the arc (x_6, x_1) belongs to the cut (X'', X'), but not to the cut (X', X'').) The capacity of the cut (X', X'') is 11.

Now let **f** be a feasible flow and let (X', X'') be a cut separating the source x_s from the sink x_t on a network G. Then summing the conservation equations (6.2a) for all those nodes x_i which belong to X', and noting cancellations, we obtain

$$\sum_{\{j|u_j\in(X'',X')\}} f_j = \sum_{\{j|u_j\in(X',X'')\}} f_j. \tag{6.26}$$

Also, since the return arc u_r belongs to (X'', X'),

$$f_r \le \sum_{\{j|u_j\in(X'',X')\}} f_j, \tag{6.27}$$

and from the feasibility condition (6.5),

$$\sum_{\{j|u_j\in(X',X'')\}} f_j \le \sum_{\{j|u_j\in(X',X'')\}} c_j. \tag{6.28}$$

Combining (6.26), (6.27), and (6.28) we obtain

$$f_r \le \sum_{\{j|u_j\in(X'',X')\}} f_j = \sum_{\{j|u_j\in(X',X'')\}} f_j \le \sum_{\{j|u_j\in(X',X'')\}} c_j. \tag{6.29}$$

Thus for any feasible flow **f**, and any cut (X', X'') separating x_s from x_t, the value of **f** is less than or equal to the capacity of (X', X''). In itself this result is hardly surprising, but it leads to the following important theorem of Ford and Fulkerson (1962):

The 'max-flow–min-cut theorem': For any network, the value of a maximal flow is equal to the minimal cut capacity of all cuts separating the source from the sink.

To prove the theorem, it suffices to show that for any given maximal flow, there exists a cut such that equality holds throughout (6.29). Indeed, let **f** be any maximal flow on G, and let $I(\mathbf{f})$ be the network obtained by removing from $\tilde{G}(\mathbf{f})$ all arcs of zero capacity, and also the arc u_r^-. (As an illustration, for the flow **f** of Fig. 6.11, the network in Fig. 6.12 is precisely the network $I(\mathbf{f})$ as defined above.) Also let X' be the set of nodes which are accessible from the source x_s on $I(\mathbf{f})$, and let $X'' = X - X'$. Since the flow **f** is maximal, $\tilde{G}(\mathbf{f})$ does not contain any flow-augmenting cycles, and consequently there are no paths from x_s to x_t on $I(\mathbf{f})$. Hence $x_s \in X'$ and $x_t \in X''$, and therefore on G, the arc set (X', X'') is a cut separating x_s from x_t.

Now from the definition of X' it follows that $I(\mathbf{f})$ does not contain any arcs with initial endpoints in X' and terminal endpoints in X''. Consequently on $\tilde{G}(\mathbf{f})$, every arc of (X', X'') other than u_r^- has a zero capacity. It follows that on G,

(i) every arc of (X'', X') other than u_r has a zero flow, which implies that equality holds in (6.27), and

(ii) every arc of (X', X'') is saturated, which implies that equality holds in (6.28).

Consequently, equality holds throughout (6.29), which proves the theorem.

Example 6.3. Let us consider the maximal flow $\mathbf{f}$ of Fig. 6.11, for which $I(\mathbf{f})$ is shown in Fig. 6.12. On Fig. 6.12, the accessible set of the source node x_1 is $\{x_1, x_2\}$. Accordingly, we partition the node set X into $X' = \{x_1, x_2\}$ and $X'' = \{x_3, x_4, x_5, x_6\}$. On Fig. 6.11, the corresponding cut (indicated by a broken line) is

$$(X', X'') = \{(x_1, x_3), (x_2, x_4)\}.$$

This cut has a capacity of 8, equal to the value of the flow. It will be observed that on Fig. 6.11, both the arcs of (X', X'') are saturated, and that the arc (x_3, x_2)—which is the only arc of (X'', X') other than the return arc (x_6, x_1)—has a zero flow.

6.5.3. An algorithm for constructing maximal flows

Let us assume that on G, the capacities of all arcs other than the return arc u_r are integers. (This is not an important restriction in practice, since a flow problem in which arcs have rational capacities can always be reduced to a problem with integer capacities, by clearing fractions.) Then a maximal flow on G can be constructed by the following algorithm.

Step 1   Choose arbitrarily some integral flow on G. (The *null flow*, in which all arc flows are zero, is a possible choice.)

Step 2   Let $\mathbf{f}$ be the present flow on G. Construct the displacement network $\tilde{G}(\mathbf{f})$, and search for a flow-augmenting cycle on this network. If no such cycle exists go to *End*.

Step 3   Let γ be a flow-augmenting cycle on $\tilde{G}(\mathbf{f})$, and let δ be the capacity of γ. For each *normal* arc u_j^+ of γ, *increase* the flow in the arc u_j on G by δ units; for each *inverted* arc u_j^- of γ, *decrease* the flow in the arc u_j of G by δ units. Then return to *Step 2*.

End   The flow on G is maximal.

It is evident that if this algorithm terminates, then the flow on G is maximal. It can also be proved that the algorithm does terminate, in a finite number of steps, by the following argument.

From the max-flow–min-cut theorem it follows that, since the capacities of all arcs other than u_r are integers, the value of a maximal flow is an integral number. Also, since the arc capacities are integers, and the computation is initiated with an integral flow, each successive flow is integral. Hence, since the flow value increases by at least one unit each time *Step 3* is executed, the algorithm constructs a flow of maximal value in a finite number of steps.

In the above proof it has emerged that for any network whose arc capacities are integers, there exists a maximal flow in which all arc flows are integers; and furthermore, our algorithm always gives a maximal flow with this property. These facts can sometimes be exploited, as in the following example.

Example 6.4. *An assignment problem* (cf. Example 5.3). A firm has q vacant jobs $\beta_1, \beta_2, \ldots, \beta_q$ of different types. There are p applicants $\alpha_1, \alpha_2, \ldots, \alpha_p$ for work with the firm, each applicant being suited for one or more of the vacancies. How should the applicants be assigned to jobs, in order to fill as many of the jobs as possible?

This problem can be solved by constructing a network in which each applicant and job is represented by a node (see Fig. 6.13). The network also

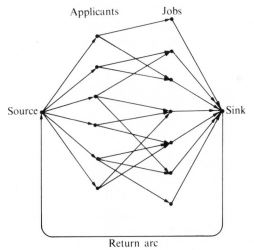

FIG. 6.13. All arcs other than the return arc have unit capacities.

contains a source node s and a sink node t. The source is joined to each applicant node α_i by an arc (s, α_i) of unit capacity; and each job node β_j is joined to the sink by an arc (β_j, t) of unit capacity. If applicant α_i is qualified for job β_j the corresponding nodes are joined by an arc (α_i, β_j), of unit capacity.

It is evident that each integral feasible flow on this network determines a feasible assignment of applicants to jobs (through the rule that, if the flow in (α_i, β_j) is unity, applicant α_i is given job β_j); a maximal integral flow defines an assignment which fills as many jobs as possible.

The successive flows and displacement networks obtained in applying the maximal flow algorithm to this problem are shown in Fig. 6.14. (For simplicity, the arc u_r^- and all arcs of zero capacity have been removed from the displacement networks.)

6.6. Minimal-cost maximal flows

6.6.1. *Minimal-cost flows*

Let us suppose that on a network G, each arc u_j is assigned a capacity c_j as before, and also a number l_j called the *unit cost* of u_j. The vector $\mathbf{l} = [l_1, l_2, \ldots, l_m]$ of these unit costs is called the *cost vector* of G. For any feasible flow $\mathbf{f}$ on G, the *total cost* of $\mathbf{f}$ is

$$\mathbf{lf}' = \sum_{j=1}^{m} l_j \cdot f_j.$$

As an illustration, in a transportation network the unit cost l_j may represent the cost of transporting one unit of a commodity along u_j, in which case the product $\mathbf{lf}'$ gives the total transportation cost incurred by the flow $\mathbf{f}$.

A feasible flow $\mathbf{f}$ on G is said to be a *minimal-cost flow* if the total cost $\mathbf{lf}'$ of $\mathbf{f}$ is less than or equal to the total cost of every other feasible flow which has the same value as $\mathbf{f}$. This section is concerned with the problem of finding a *minimal-cost maximal flow*, that is, a maximal flow whose total cost is as small as possible.

6.6.2. *Costs on displacement networks*

For any feasible flow $\mathbf{f}$ we assign unit costs to the arcs of $\tilde{G}(\mathbf{f})$ as follows. Each *normal* arc u_j^+ of $\tilde{G}(\mathbf{f})$ is given the unit cost l_j of u_j, and each *inverted* arc u_j^- of $\tilde{G}(f)$ is assigned the unit cost $-l_j$. Thus the cost vector $\mathbf{\tilde{l}}$ of $\tilde{G}(\mathbf{f})$ can be written as

$$\mathbf{\tilde{l}} = [\mathbf{l} \mid -\mathbf{l}]. \tag{6.30}$$

Successive network flows
The bold lines indicate arcs carrying one unit of flow; all other arcs (except the return arc) have zero flows.

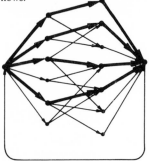

Flow value = 4

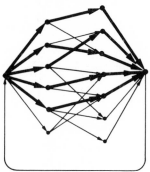

Flow value = 5

Flow value = 6

Displacement networks
Arcs with zero capacities and the arc u_r^- have been omitted. The bold lines indicate flow-augmenting cycles.

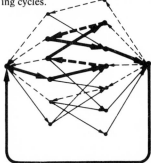

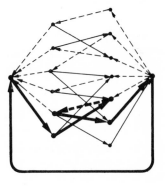

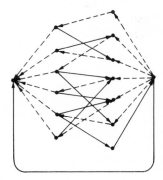

FIG. 6.14

From (6.30) it follows that the total cost of a feasible flow $\tilde{\mathbf{f}} = [\mathbf{f}^+ | \mathbf{f}^-]$ on $\hat{G}(\mathbf{f})$ can be expressed as

$$\mathbf{lf}' = \mathbf{l}(\mathbf{f}^+ - \mathbf{f}^-)'. \tag{6.31}$$

Now in section 6.4 it was shown that for any two feasible flows $\mathbf{f}$ and $\mathbf{g}$ on G, the flow difference $\mathbf{g} - \mathbf{f}$ determines a disjunctive feasible flow $\tilde{\mathbf{f}} = [\mathbf{f}^+ | \mathbf{f}^-]$ on $\tilde{G}(\mathbf{f})$, such that (cf. (6.17))

$$\mathbf{f}^+ - \mathbf{f}^- = \mathbf{g} - \mathbf{f}. \tag{6.32}$$

From (6.31) and (6.32), it follows immediately that

$$\widetilde{\mathbf{lf}}' = \mathbf{lg}' - \mathbf{lf}'. \tag{6.33}$$

Thus, the difference between the total costs of $\mathbf{f}$ and $\mathbf{g}$ is given directly by the cost of $\tilde{\mathbf{f}}$.

Conversely, for any feasible flow $\mathbf{f}$ on G, and any feasible flow $\tilde{\mathbf{f}} = [\mathbf{f}^+ | \mathbf{f}^-]$ on $\tilde{G}(\mathbf{f})$, the flow $\mathbf{g}$ on G defined by (6.18), viz.

$$\mathbf{g} = \mathbf{f} + \mathbf{f}^+ - \mathbf{f}^-,$$

has a total cost

$$\mathbf{lg}' = \mathbf{l}(\mathbf{f} + \mathbf{f}^+ - \mathbf{f}^-)' = \mathbf{lf}' + \mathbf{l}(\mathbf{f}^+ - \mathbf{f}^-)' = \mathbf{lf}' + \widetilde{\mathbf{lf}}'. \tag{6.34}$$

Thus, the total cost of $\mathbf{g}$ is the sum of the total costs of $\mathbf{f}$ and $\tilde{\mathbf{f}}$.

As a consequence of these results it is possible to determine whether a given flow $\mathbf{f}$ is of minimal cost by inspection of $\tilde{G}(\mathbf{f})$, in the manner described below.

6.6.3. Cost-reducing cycles

Let $\mathbf{f}$ be any feasible (but not necessarily maximal) flow on G, and let us suppose that $\tilde{G}(\mathbf{f})$ contains an elementary cycle γ with the following properties:

(i) γ does not contain either of the arcs u_r^+ or u_r^- ;
(ii) all arcs of γ have non-zero capacities; and
(iii) the sum of the unit costs of the arcs of γ is negative.

Let δ be the smallest of the arc capacities on γ, and let $\tilde{\mathbf{f}} = [\mathbf{f}^+ | \mathbf{f}^-]$ be the feasible flow on $\tilde{G}(\mathbf{f})$ obtained by assigning a flow of δ units to each arc of γ.

Now consider the flow

$$\mathbf{g} = \mathbf{f} + \mathbf{f}^+ - \mathbf{f}^-$$

on G. Since γ has property (i), $f_r^+ = 0$ and $f_r^- = 0$, hence the flow **g** has the same value as **f**. Also, since γ has property (iii) the total cost of $\tilde{\mathbf{f}}$ is negative,

$$\tilde{\mathbf{l}}\tilde{\mathbf{f}}' < 0. \tag{6.35}$$

It follows from (6.34) and (6.35) that

$$\mathbf{l}\mathbf{g}' = \mathbf{l}\mathbf{f}' + \tilde{\mathbf{l}}\tilde{\mathbf{f}}' < \mathbf{l}\mathbf{f}'. \tag{6.36}$$

Thus, the flow **g** has the same value as **f**, but a smaller total cost. For obvious reasons, we describe any elementary cycle γ on $\tilde{G}(\mathbf{f})$ which has the properties (i)–(iii) as a *cost-reducing cycle*.

From the above it is clear that if a displacement network $\tilde{G}(\mathbf{f})$ contains a cost-reducing cycle, then the flow **f** is not of minimal cost. Conversely, it can be proved that if a flow **f** is not of minimal cost, then $\tilde{G}(\mathbf{f})$ contains at least one cost-reducing cycle. Indeed, let **f** and **g** be two feasible flows of the same value, and let us suppose that $\mathbf{l}\mathbf{f}' > \mathbf{l}\mathbf{g}'$. Let $\tilde{\mathbf{f}}$ be the disjunctive flow on $\tilde{G}(\mathbf{f})$ determined by $\mathbf{g} - \mathbf{f}$. Then since $\mathbf{l}\mathbf{f}' > \mathbf{l}\mathbf{g}'$, it follows from (6.33) that the total cost of $\tilde{\mathbf{f}}$ is negative,

$$\tilde{\mathbf{l}}\tilde{\mathbf{f}}' < 0. \tag{6.37}$$

Now $\tilde{\mathbf{f}}$ can be expressed in the form

$$\tilde{\mathbf{f}} = r_1\tilde{\mathbf{v}}_1 + r_2\tilde{\mathbf{v}}_2 + \cdots + r_k\tilde{\mathbf{v}}_k, \tag{6.38}$$

where $r_1, r_2, \ldots, r_k$ are positive numbers and $\tilde{\mathbf{v}}_1, \tilde{\mathbf{v}}_2, \ldots, \tilde{\mathbf{v}}_k$ are elementary cyclic flows on $\tilde{G}(\mathbf{f})$. Since **f** and **g** have the same value, $f_r^+ = 0$ and $f_r^- = 0$, and consequently all the cycles $\gamma_1, \gamma_2, \ldots, \gamma_k$ associated with the flows $\tilde{\mathbf{v}}_1, \tilde{\mathbf{v}}_2, \ldots, \tilde{\mathbf{v}}_k$ have property (i) above. Also, since all the numbers $r_1, r_2, \ldots, r_k$ in (6.38) are non-zero, the cycles $\gamma_1, \gamma_2, \ldots, \gamma_k$ all have property (ii). Finally, it follows from (6.37) that for some elementary flow $\tilde{\mathbf{v}}_i$ say in (6.38), $\tilde{\mathbf{l}}\tilde{\mathbf{v}}_i' < 0$, which implies that γ_i has property (iii). Hence $\tilde{G}(\mathbf{f})$ contains a cost-reducing cycle, as required.

Combining the results of this section, we obtain the following characterization of minimal-cost flows:

A flow **f** *on G is a minimal-cost flow if and only if the displacement network $\tilde{G}(\mathbf{f})$ does not contain any cost-reducing cycles.*

6.6.4. Algorithms for constructing minimal-cost maximal flows

The cost-reduction method. The characterization of minimal-cost flows of the previous section immediately suggests the following algorithm:

Step 1   Construct a maximal flow on G. (The algorithm of Section 6.5.3 can be used for this purpose.)

Step 2   Let **f** be the present flow on G. Construct the displacement network $\tilde{G}(\mathbf{f})$, and search for a cost-reducing cycle on this network. If no such cycle exists go to *End*.

Step 3   Let γ be a cost-reducing cycle on $\tilde{G}(\mathbf{f})$, and let δ be the smallest of its arc capacities. Modify the present flow on G as follows: For each *normal* arc u_j^+ of γ, *increase* the flow in the corresponding arc u_j on G by δ units, and for each *inverted* arc u_j^- of γ, *decrease* the flow in the corresponding arc u_j by δ units. Then return to *Step 2*.

End   The flow on G is a minimal-cost maximal flow.

As an illustration, for the network of Fig. 6.15, the successive flows and displacement networks obtained by the cost-reduction method are shown in Fig. 6.16. (For simplicity the arcs u_r^+, u_r^-, and all arcs of zero capacity have been removed from the displacement networks.)

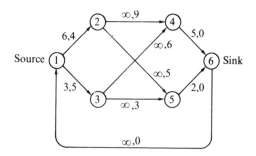

FIG. 6.15. The numbers indicate arc capacities and costs.

The flow-augmentation method. In the previous method we obtained a minimal-cost maximal flow by constructing a sequence of maximal flows of successively smaller costs, until the flow cost was as small as possible. Alternatively, if the unit costs of all the arcs of a network are non-negative, we can construct a sequence of minimal-cost flows with successively greater values, until the flow value becomes maximal. It is very easy to obtain an appropriate initial flow for this method: for if the unit costs of all arcs are

Successive network flows
The numbers indicate capacities and flows.

Displacement networks
The numbers indicate capacities and unit costs; bold lines indicate cost-reducing cycles.

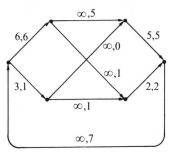

Flow cost $=82$

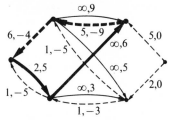

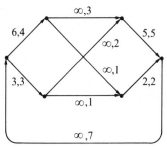

Flow cost $=78$

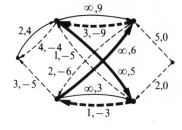

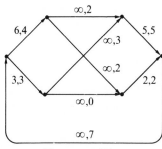

Flow cost $=77$

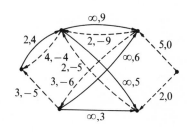

FIG. 6.16

non-negative, the null flow $f=0$ is obviously a minimal-cost flow of value zero! The method of constructing the subsequent flows is defined and justified in the following theorem:

Let f be a minimal-cost flow on G, let γ be a flow-augmenting cycle of minimal cost on $\tilde{G}(f)$, and let δ be the smallest arc capacity of γ.

Then the vector **g** *defined by*

$$
g_i = \begin{cases} f_i + \delta & \text{if the normal arc } u_i^+ \text{ belongs to } \gamma, \\ f_i - \delta & \text{if the reverse arc } u_i^- \text{ belongs to } \gamma, \\ f_i & \text{if neither } u_i^+ \text{ nor } u_i^- \text{ belongs to } \gamma, \end{cases} \quad (6.39)
$$

is a minimal-cost feasible flow on G, whose value is δ units greater than the value of **f**.

Indeed, from the results of Section 6.5.1 it is evident that **g** is a feasible flow, of value $g_r = f_r + \delta$. To prove that **g** is of minimal cost we compare its cost with that of any other flow **h** which has the same value. For this purpose, let us denote by $\tilde{\mathbf{f}}_g$ and $\tilde{\mathbf{f}}_h$ the disjunctive flows on $\tilde{G}(\mathbf{f})$ determined by the flow differences $\mathbf{g} - \mathbf{f}$ and $\mathbf{h} - \mathbf{f}$ respectively. The total costs of these flows are

$$
\tilde{\mathbf{l}}\tilde{\mathbf{f}}_g' = \mathbf{l}\mathbf{g}' - \mathbf{l}\mathbf{f}' \quad (6.40)
$$

and

$$
\tilde{\mathbf{l}}\tilde{\mathbf{f}}_h' = \mathbf{l}\mathbf{h}' - \mathbf{l}\mathbf{f}'. \quad (6.41)
$$

Now the flow $\tilde{\mathbf{f}}_g$ can be expressed in the form

$$
\tilde{\mathbf{f}}_g = \delta \tilde{\mathbf{v}} \quad (6.42)
$$

where $\tilde{\mathbf{v}}$ is the flow on $\tilde{G}(\mathbf{f})$ obtained by assigning one unit of flow to each arc of γ. Also, the flow $\tilde{\mathbf{f}}_h$ can be expressed in terms of elementary cyclic flows:

$$
\tilde{\mathbf{f}}_h = \sum_{i=1}^{j} a_i \tilde{\mathbf{p}}_i + \sum_{i=1}^{k} b_i \tilde{\mathbf{q}}_i, \quad (6.43)
$$

where $a_1, a_2, \ldots a_j$ and $b_1, b_2, \ldots, b_k$ are positive numbers, $\tilde{\mathbf{p}}_1, \tilde{\mathbf{p}}_2, \ldots, \tilde{\mathbf{p}}_j$ are elementary flows associated with flow-augmenting cycles, and $\tilde{\mathbf{q}}_1, \tilde{\mathbf{q}}_2, \ldots, \tilde{\mathbf{q}}_k$ are elementary flows associated with cycles which contain neither u_r^+ nor u_r^-. The value of **h** is δ units greater than the value of **f**, therefore

$$
\sum_{i=1}^{j} a_i = \delta. \quad (6.44)
$$

Since the flow-augmenting cycle γ is of minimal cost,

$$
\tilde{\mathbf{l}}\tilde{\mathbf{v}}' \leq \tilde{\mathbf{l}}\tilde{\mathbf{p}}_i', \qquad (i = 1, 2, \ldots, j). \quad (6.45)
$$

Hence, from (6.42), (6.44), and (6.45),

$$\tilde{\mathbf{lf}}'_g = \delta\tilde{\mathbf{lv}}' \leq \sum_{i=1}^{j} a_i \tilde{\mathbf{lp}}'_i. \tag{6.46}$$

Also, since $\tilde{\mathbf{f}}$ is a minimal-cost flow, $\tilde{G}(\mathbf{f})$ does not contain any cost-reducing cycles; consequently

$$\tilde{\mathbf{lq}}'_i \geq 0, \qquad (i = 1, 2, \ldots, k), \tag{6.47}$$

and therefore

$$\sum_{i=1}^{k} b_i \tilde{\mathbf{lq}}'_i \geq 0. \tag{6.48}$$

Now from (6.43), the cost of $\tilde{\mathbf{f}}_h$ can be expressed as

$$\tilde{\mathbf{lf}}'_h = \sum_{i=1}^{j} a_i \tilde{\mathbf{lp}}'_i + \sum_{i=1}^{k} b_i \tilde{\mathbf{lq}}'_i \tag{6.49}$$

and combining (6.46), (6.48) and (6.49) we obtain

$$\tilde{\mathbf{lf}}'_g \leq \tilde{\mathbf{lf}}'_h. \tag{6.50}$$

Consequently, from (6.40), (6.41), and (6.50),

$$\mathbf{lg}' \leq \mathbf{lh}'.$$

The flow $\mathbf{g}$ is therefore of minimal cost, as required.

Thus, in the flow-augmentation method, we follow precisely the procedure defined in Section 6.5.3 for constructing a maximal flow, with the understanding that

(i) in *Step 1* the initial flow is a minimal-cost flow, and
(ii) in *Step 2* and *Step 3*, the required flow-augmenting cycle γ is of minimal cost.

For the network of Fig. 6.15, the successive flows obtained by this method are shown in Fig. 6.17.

6.7. Transportation and assignment problems

We shall now give some examples of minimal-cost maximal flow problems, which arise frequently in operational research.

6.7.1. *The transportation problem*

This problem, which is perhaps the most important instance of a network-flow problem, arises for example in transporting coal from

Successive displacement networks
The numbers indicate arc capacities and unit costs; bold lines indicate minimal-cost flow-augmenting cycles.

Successive network flows
The numbers indicate arc capacities and flows; bold lines indicate saturated arcs.

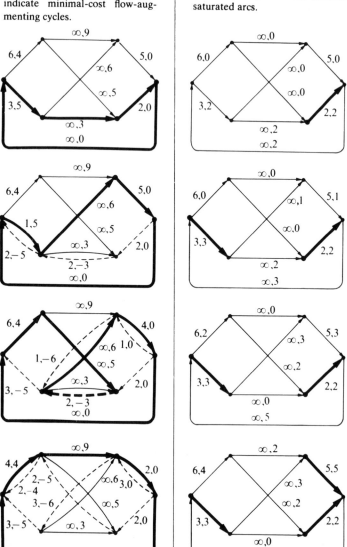

Fig. 6.17

coalfields to power stations. The problem can be stated in general terms as follows: There are p sources $\alpha_1, \alpha_2, \ldots, \alpha_p$ of a commodity, with c_i units of supply at each node α_i. The commodity is required at q destinations $\beta_1, \beta_2, \ldots, \beta_q$, the demand at each destination β_j being for d_j units. The cost of sending one unit from each source α_i to each destination β_j is λ_{ij}; it can be assumed that the amount of the commodity which can be carried from each origin to each destination is unlimited. How should the commodity be distributed, from the sources to the destinations, in order to meet all the demands at minimum total cost?

This problem can be considered as a minimal-cost maximal flow problem on a network of the form shown in Fig. 6.18, in which the first number on each arc represents its capacity and the second is its unit cost.

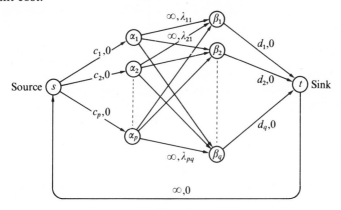

FIG. 6.18

6.7.2. *The optimal assignment problem*

Let us suppose that in a machine-shop, there are q jobs $\beta_1, \beta_2, \ldots, \beta_q$ to be executed concurrently by different machines, and p machines $\alpha_1, \alpha_2, \ldots, \alpha_p$ available to execute these jobs. The machines vary in their suitability for a particular job; we suppose that these variations can be represented by attributing to each possible assignment $\alpha_i \to \beta_j$ a *cost* λ_{ij}, which is a real number. An *optimal assignment* is one in which machines are assigned to all the jobs, at minimum total cost.

The problem of constructing an optimal assignment can be considered as a particular case of the transportation problem presented above: we may regard the machines as sources of a commodity (each capable of supplying one unit), the jobs as consumers (each requiring one unit), and consider the assignment of machine α_i to job β_j as a shipment of one unit of the commodity from α_i to β_j. Thus the assignment problem can be regarded as the problem of finding a minimal-cost integral maximal flow on a network of the form shown in Fig. 6.18, in which all the arc capacities c_i and d_j are set to unity.

There are many variants of these transportation and assignment problems which at first sight appear to be more complicated, but which reduce to flow problems on networks of the same basic form; the following is an example.

6.7.3. *The assignment of machines to a fixed schedule of tasks*

Let us suppose that it is necessary to execute q tasks $\beta_1, \beta_2, \ldots, \beta_q$, where each task β_j has a stipulated starting time σ_j and finishing time ϕ_j, and where each task requires one machine. To perform these tasks we have p machines $\alpha_1, \alpha_2, \ldots, \alpha_p$, which can be considered to be identical. A machine can only perform one task at a time, but can execute any number of tasks in succession.

Before a machine can perform any particular task, it must be 'set up' or adjusted for the purpose. The time taken and the cost of setting-up the machine depend on the nature of the task, and on what the machine was doing previously; to set up machine α_i (from its initial state) for task β_j, the time required is δ_{ij}, and the cost λ_{ij}. To 'reset' the machine which performed task β_j for task β_k takes a time δ'_{jk}, and the cost of this re-set operation is λ'_{jk}. It is required to assign the machines to tasks in such a way that all tasks are performed, at minimal total cost.

As an illustration, the tasks β_j could be scheduled airline flights between different cities, and the machines α_i aircraft all of the same type. In this case δ_{ij} and λ_{ij} would be the time required and the cost of flying the aircraft α_i from its initial location to the departure point of the flight β_j; whereas δ'_{jk} and λ'_{jk} would represent the time and cost of flying *any* aircraft from the arrival point of flight β_j to the departure point of flight β_k.

The problem can be reduced to an optimal assignment problem of the type considered in the previous section: we take as our set of machines the p machines $\alpha_1, \alpha_2, \ldots, \alpha_p$ which are initially available, together with q additional machines $\gamma_1, \gamma_2, \ldots, \gamma_q$, where γ_j represents the machine which performs task β_j (and which becomes available for assignment to other tasks when this task has been completed).

A machine α_i is considered capable of performing a task β_j if and only if $\delta_{ij} \le \sigma_j$, and the cost of each feasible assignment $\alpha_i \to \beta_j$ is λ_{ij}. A machine γ_j is considered capable of performing task β_k if and only if $\phi_j + \delta'_{jk} \ge \sigma_k$, and the cost of each feasible assignment $\gamma_j \to \beta_k$ is λ'_{jk}.

A network model of this assignment problem is shown in Fig. 6.19. If the maximal flows on this network saturate all the arcs

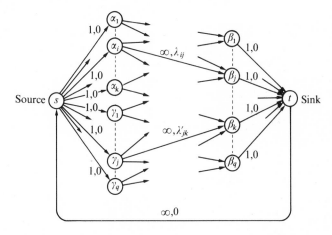

FIG. 6.19. The numbers indicate arc capacities and costs.

(β_j, t), then it is possible to execute all the tasks. A minimal-cost integral maximal flow then defines an optimal assignment, through the following rules:

(i) if the flow in an arc (α_i, β_j) is unity then machine α_i is initially assigned to the task β_j, and
(ii) if the flow in arc (γ_j, β_k) is unity then the machine which performs task β_j is next assigned the task β_k.

Several other examples of transportation and assignment problems are described in the Exercises at the end of this chapter.

6.8. Circulations

6.8.1. Definition of a circulation

In the preceding sections it has been assumed that a network contains a source and a sink, joined by a return arc of infinite capacity. Here we no longer distinguish a source node or sink node. Instead we suppose that for each arc u_j of a network we have a specified *lower bound* $b_j \geq 0$ on the arc flow, as well as an upper bound (or capacity) c_j.

In this case, a flow $\mathbf{f} = [f_1, f_2, \ldots, f_m]$ on a network G is called a *circulation* if

$$\mathbf{b} \leq \mathbf{f} \leq \mathbf{c}, \tag{6.51}$$

where $\mathbf{b} = [b_1, b_2, \ldots, b_m]$ is the vector of lower bounds on arc flows and $\mathbf{c} = [c_1, c_2, \ldots, c_m]$ is the capacity vector of G.

The first problem to be considered here is that of finding a circulation on G—if any circulations exist. (It is evident that there may not be any flow $\mathbf{f}$ which satisfies the condition (6.51).) It will then be shown how to construct a circulation of minimal cost.

6.8.2. Auxiliary networks

The problem of constructing a circulation on a network G can be reduced to a maximal flow problem on a modified network C^0, in which the lower bounds on arc flows are all zero. This network G^0, called the *auxiliary network* of G, is obtained as follows.

The node set of G^0 consists of the nodes $x_1, x_2, \ldots, x_n$ of G, together with a source node s and a sink node t. The arc set of G^0 comprises the following:

(i) For every arc u_j of G, the network G^0 contains an arc v_j^U which has the same initial and terminal endpoints as u_j, and a capacity

$$c_j^U = c_j - b_j, \qquad (j = 1, 2, \ldots, m). \tag{6.52}$$

(ii) For every node x_i of G, the network G^0 cpntains an *entry arc* v_i^S with initial endpoint s and terminal endpoint x_i, and also an *exit arc* v_i^T with initial endpoint x_i and terminal endpoint t. The capacities of the arcs v_i^S and v_i^T are

$$c_i^S = \max\{0, d_i\} \quad \text{and} \quad c_i^T = \max\{0, -d_i\} \tag{6.53(a)}$$

respectively, where

$$d_i = \sum_{\{j|s_{ij}=-1\}} b_j - \sum_{\{j|s_{ij}=+1\}} b_j, \qquad (i = 1, 2, \ldots, n).$$

(6.53(b))

(iii) G^0 contains a *return arc* v^R from t to s, which has a capacity

$$c^R = \infty.$$

(6.54)

It will be noted that in the right-hand side of (6.53(b)), the first term represents the sum of the lower bounds on all the arcs incident to x_i, while the second term is the sum of the lower bounds on the arcs incident from x_i. If the difference d_i between these sums is positive, the entry arc v_i^S has a capacity $c_i^S = d_i$ and the exit arc v_i^T has a zero capacity; whereas if d_i is negative, the exit arc v_i^T has a capacity $c_i^T = -d_i$ and the entry arc c_i^S has zero capacity. In practical computations, entry or exit arcs of zero capacity can be omitted; we retain them only to simplify our algebraic presentation.

As an illustration, the network of Fig. 6.20 has the auxiliary network shown in Fig. 6.21.

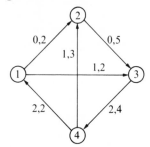

FIG. 6.20. The numbers indicate lower bounds and capacities.

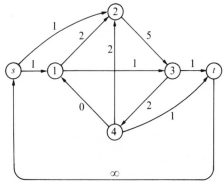

FIG. 6.21. The numbers indicate arc capacities.

Let us suppose that the nodes of G are ordered as follows: $s, t, x_1, x_2, \ldots, x_n$, and that its arcs are ordered as follows:

$$v^R, \underbrace{v_1^S, v_2^S, \ldots, v_n^S}_{\substack{\text{Entry} \\ \text{arcs}}}, \underbrace{v_1^T, v_2^T, \ldots, v_n^T}_{\substack{\text{Exit} \\ \text{arcs}}}, \underbrace{v_1^U, v_2^U, \ldots, v_m^U}_{\substack{\text{Arcs} \\ \text{of } G}}.$$

with labels: Return arc (v^R), Entry arcs, Exit arcs, Arcs of G.

Then the node-arc incidence matrix S^0 of G^0 can be written in the partitioned form

$$S^0 = \begin{bmatrix} -1 & J & 0 & 0 \\ +1 & 0 & -J & 0 \\ 0 & -I & I & S \end{bmatrix}, \qquad (6.55)$$

where I is an $n \times n$ unit matrix, J is the *universal* row vector whose n elements all have the value 1, and S is the incidence matrix of G.

With a corresponding partitioning, the capacity vector $\mathbf{c}^0$ of G^0 has the form

$$\mathbf{c}^0 = [\mathbf{c}^R | \mathbf{c}^S | \mathbf{c}^T | \mathbf{c}^U] \qquad (6.56)$$

where $\mathbf{c}^R$ is the capacity of the return arc, $\mathbf{c}^S$ is the vector of capacities of the entry arcs, $\mathbf{c}^T$ is the vector of capacities of the exit arcs, and $\mathbf{c}^U$ is the vector of capacities of those arcs of G^0 which appear also in G.

In vector form, the condition (6.52) can be written as

$$\mathbf{c}^U = \mathbf{c} - \mathbf{b}. \qquad (6.57)$$

Also, the vector $\mathbf{d} = [d_1, d_2, \ldots, d_n]$ of elements d_i defined by (6.53(b)) can be expressed as

$$\mathbf{d} = -\mathbf{b}S'$$

and therefore, by (6.53(a)), the vectors $\mathbf{c}^S$ and $\mathbf{c}^T$ satisfy the condition

$$\mathbf{c}^S - \mathbf{c}^T = -\mathbf{b}S'. \qquad (6.58)$$

Transposing both sides of (6.58), and premultiplying by the universal vector J we obtain

$$J(\mathbf{c}^S - \mathbf{c}^T)' = JS\mathbf{b}' \qquad (6.59)$$

and since $JS = 0$ (as a consequence of the fact that each column of S has precisely two non-zero elements, with values of $+1$ and -1), the equation (6.59) reduces to

$$J(\mathbf{c}^S)' = J(\mathbf{c}^T)'. \tag{6.60}$$

In words, (6.60) states that the sum of the capacities of the entry arcs is equal to the sum of the capacities of the exit arcs.

Now let $\mathbf{f}^0$ be any feasible flow on G^0. If $\mathbf{f}^0$ is partitioned in the same manner as $\mathbf{c}^0$, viz.

$$\mathbf{f}^0 = [\mathbf{f}^R | \mathbf{f}^S | \mathbf{f}^T | \mathbf{f}^U] \tag{6.61}$$

then the flow feasibility condition $0 \le \mathbf{f}^0 \le \mathbf{c}^0$ can be written as

$$0 \le \mathbf{f}^R \le \mathbf{c}^R, \tag{6.62(a)}$$

$$0 \le \mathbf{f}^S \le \mathbf{c}^S, \tag{6.62(b)}$$

$$0 \le \mathbf{f}^T \le \mathbf{c}^T, \tag{6.62(c)}$$

$$0 \le \mathbf{f}^U \le \mathbf{c}^U. \tag{6.62(d)}$$

Also, from (6.55) and (6.61), the flow conservation condition $S^0(\mathbf{f}^0)' = 0$ can be expressed as

$$J(\mathbf{f}^S)' = \mathbf{f}^R = J(\mathbf{f}^T)', \tag{6.63(a)}$$

$$S(\mathbf{f}^U)' = (\mathbf{f}^S - \mathbf{f}^T)'. \tag{6.63(b)}$$

In words, (6.63(a)) states that the sum of the flows in the entry arcs is equal to the flow in the return arc, and that this flow is also equal to the sum of the flows in the exit arcs. We observe that (6.63(a)) and (6.60) together imply that

$$\mathbf{f}^S = \mathbf{c}^S \quad \textit{if and only if} \quad \mathbf{f}^T = \mathbf{c}^T; \tag{6.64}$$

we shall describe a flow which saturates all the entry and exit arcs of G^0 as a *saturating flow*.

6.8.3. The construction of circulations

Let us suppose that there exists a saturating flow on G^0, and let $\mathbf{f}^0 = [\mathbf{f}^R | \mathbf{f}^S | \mathbf{f}^T | \mathbf{f}^U]$ be such a flow. Also, let $\mathbf{f}$ be the vector of m elements defined by

$$\mathbf{f} = \mathbf{f}^U + \mathbf{b}. \tag{6.65}$$

From (6.57) and (6.62(d)),

$$0 \le \mathbf{f}^U \le \mathbf{c} - \mathbf{b},$$

hence by (6.65)

$$\mathbf{b} \le \mathbf{f} \le \mathbf{c}. \qquad (6.66)$$

Also, it follows from (6.65), (6.63(b)) and (6.58) that

$$\mathbf{Sf'} = S(\mathbf{f}^U)' + \mathbf{Sb'} \qquad (6.67)$$
$$= (\mathbf{f}^S - \mathbf{f}^T)' - (\mathbf{c}^S - \mathbf{c}^T)'$$
$$= (\mathbf{f}^S - \mathbf{c}^S)' - (\mathbf{f}^T - \mathbf{c}^T)'$$
$$= 0. \qquad (6.68)$$

From (6.66) and (6.68), it follows that $\mathbf{f}$ is a circulation on G.

Conversely, it is evident that each circulation $\mathbf{f}$ on G determines a saturating flow $\mathbf{f}^0 = [\mathbf{f}^R | \mathbf{f}^S | \mathbf{f}^T | \mathbf{f}^U]$ on G^0 through the rule (6.65), which gives

$$\mathbf{f}^R = J(\mathbf{c}^S)', \qquad \mathbf{f}^S = \mathbf{c}^S, \qquad \mathbf{f}^T = \mathbf{c}^T, \qquad \mathbf{f}^U = \mathbf{f} - \mathbf{b}. \qquad (6.69)$$

As an illustration, Fig. 6.22 shows a circulation on the network of Fig. 6.20; the corresponding saturating flow on G^0 is shown in Fig. 6.23.

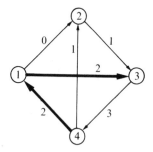

FIG. 6.22. The numbers indicate arc flows; bold lines indicate saturated arcs.

From these arguments, it follows that we can obtain a circulation on a network G simply by applying the maximum flow algorithm of Section 6.5.3 to the corresponding auxiliary network G^0: if this produces a saturating flow on G^0, the corresponding circulation is given immediately by (6.65); whereas if the maximal flow obtained on G^0 is not a saturating flow, then no circulations exist for G.

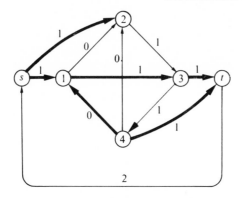

FIG. 6.23. The numbers indicate arc flows; bold lines indicate saturated arcs.

6.8.4. *Minimal-cost circulations*

Let us suppose that on G^0, each arc v_j^U is assigned the unit cost of the corresponding arc u_j on G:

$$l_j^U = l_j, \qquad (j = 1, 2, \ldots, m) \tag{6.70}$$

and that all other arcs of G^0 are assigned unit costs of zero. Then a saturating flow f^0 on G^0 has a total cost

$$l^0(f^0)' = l(f^U)'. \tag{6.71}$$

The total cost of the corresponding circulation $f = f^U + b$ on G is

$$lf' = l(f^U)' + lb'. \tag{6.72}$$

From (6.71) and (6.72), the total costs of f and f^0 are related by

$$lf' = l^0(f^0)' + lb'. \tag{6.73}$$

Hence if f^0 is a minimal-cost saturating flow on G^0, then f is a minimal-cost circulation on G. Consequently we can obtain a minimal-cost circulation on G simply by constructing a minimal-cost maximal flow on G^0, using one of the algorithms of Section 6.6.4 and then applying (6.65).

Example 6.5. *Production planning.* A manufacturing firm has to meet a demand for one of its products over k successive time-periods, the numbers of units to be despatched to customers in each period being $d_1, d_2, \ldots, d_k$.

Production costs vary from one period to the next, because of changes in costs of raw materials: the expected unit production costs for each period are $l_1, l_2, \ldots, l_k$. The firm must make at least m units in each period, but up to n additional units can be made by overtime working, at an additional unit cost of e. Up to c units can be stored in a warehouse, the unit cost of storage from one period to the next being w. How many units should be manufactured in each period, to meet the demands at minimum total cost?

A network model of this problem is shown in Fig. 6.24 (for the case where $k = 4$). It is evident that a minimal-cost circulation on this network determines an optimal production pattern, the flows in the arcs joining node s to node p_i being the numbers of units to be manufactured by normal and overtime working in the ith period.

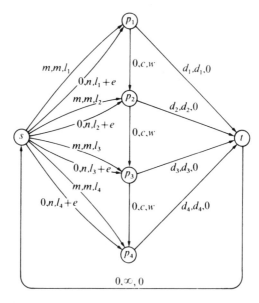

FIG. 6.24. The labels on arcs are their lower flow bounds, capacities, and unit costs.

6.9. Practical considerations

We shall now consider briefly the practical aspects of the application of the algorithms of Section 6.6.4 for constructing a minimal-cost maximal flow.

6.9.1. Implementation of the flow-augmentation method

In practice, instead of constructing a displacement network $\tilde{G}(\mathbf{f})$ and searching for a flow-augmenting cycle on this network, it is convenient to construct the network $I(\mathbf{f})$, which only differs from $\tilde{G}(\mathbf{f})$ in that the arcs of zero capacity and the inverted arc u_r^- are omitted (see for example Fig. 6.17). Then, if the arc costs on $I(\mathbf{f})$ are regarded as 'lengths', a shortest path from the source s to the sink t on this graph, together with the arc u_r^+, constitutes a minimal-cost flow-augmenting cycle. Since the flow $\mathbf{f}$ constructed at each stage of the flow-augmentation method is a minimal-cost flow, the network $I(\mathbf{f})$ does not contain any cycles of negative length. The required shortest path from s to t can therefore be obtained by any of the direct methods of Section 3.6 (such as Gauss elimination), or the iterative methods of Section 3.7 (such as Yen's double-sweep method). In this application, Yen's method is usually the most efficient.

As an alternative strategy, Edmonds and Karp (1972) and Tomizawa (1971) have shown that it is easily possible to modify the arc costs on $I(\mathbf{f})$ in such a way that they all become non-negative, and yet the paths from s to t which are shortest on $I(\mathbf{f})$ remain the shortest paths between these nodes on the modified network. With all arc costs non-negative, it becomes possible to construct a shortest path from s to t by Dijkstra's method (see Section 3.8), which is only of complexity $O(n^2)$—or even $O(n \log n)$ if one employs a sort tree—as compared with the complexity $O(n^3)$ of the direct and iterative methods mentioned above. However, Edmonds and Karp carefully describe their technique as giving a *theoretical* improvement in algorithmic efficiency: an improvement in the upper bound on running times does not imply a better practical performance. In some experimental comparisons made by the author of this book, for relatively small but realistic problems, the use of Yen's method on $I(\mathbf{f})$ always gave the best results.

With regard to the number of flow augmentations performed by the algorithm, it is evident that if all arc capacities are integers then the total number of augmentations is not greater than the value of a maximal flow. If at any stage a displacement network $\tilde{G}(\mathbf{f})$ contains several minimal-cost flow-augmenting cycles, intuitively it would seem possible that, by choosing among these a cycle of lowest order, we might reduce the total number of augmentations required. A theoretical justification of this refinement is given by Edmonds and

Karp (1972). The refinement is very easily incorporated in a path-finding algorithm: if all the arc lengths are integers, we simply change each arc length l_j to $kl_j + 1$, where k is any constant greater than the total number of arcs.

In applying the flow-augmentation method, it is sometimes possible to exploit particular characteristics of a problem. For instance in the case of transportation problems of the kind described in Section 6.7, it is possible to modify the arc costs in such a way as to facilitate the choice of an initial flow of value greater than zero (see Exercise 6.7). For the same problem, it is also possible to simplify the displacement networks, by removing many of their inverted arcs (see Exercise 6.8). Other refinements of the method, for the solution of assignment problems, are described by Tabourier (1972).

Finally, we must draw attention to the 'primal–dual' method of Ford and Fulkerson (1962), whose development was motivated by considerations of duality in linear programming. In essence this is very similar to the flow-augmentation method presented here, but each successive flow is found by solving a maximal flow problem on a partial network of the original one.

6.9.2. Implementation of the cost-reduction method

The cost-reducing cycles on a displacement network $\tilde{G}(\mathbf{f})$ correspond to the cycles of negative length on the network $I(\mathbf{f})$, as defined above; a technique for finding such a cycle has already been presented in Chapter 3 (see Exercise 3.10 and its solution). A technique of this kind is used in the network flow algorithm of Klein (1967).

For transportation problems in particular a special form of cost-reduction method has been devised, in which the determination of the cost-reducing cycles is greatly simplified, through a judicious choice of the initial flow. This method, which is widely known as the 'stepping-stone method', was originally developed by G. B. Dantzig as a particular form of his Simplex method for solving linear programs. For a description of this method in graph-theoretic terms see Dantzig (1963), Chapters 14–17.

6.9.3. Comparison of methods

With most network flow problems, the determination of a least-cost flow-augmenting cycle involves much less work than the determination of a cost-reducing cycle. A flow-augmentation method is therefore likely to be more efficient than a cost-reduction method, unless a maximal flow of almost minimal cost is available initially.

For transportation problems in particular, a great deal of effort has been devoted to the development of efficient programs, these being based mostly on the Ford–Fulkerson primal–dual method and Dantzig's Simplex method, and comparative results have been published frequently. Although the conclusions drawn by different authors on the relative merits of the primal–dual and Simplex methods have often been contradictory, it would seem that programs based on the primal–dual method have a better performance (Hatch 1975). Some comparisons between the basic flow-augmentation method (using Gauss elimination to find flow-augmenting paths) and the primal–dual method suggest that the former is better for problems of modest size, whereas the primal–dual method is more efficient for large ones (Carré 1971).

Exercises

6.1. *Node capacities.* Let $G = (X, U)$ be a network in which each *node* $x_i \in X$ has a flow capacity $k_i \geq 0$, this being an upper limit to the sum of flows in arcs incident to (or from) node x_i. Show that the problem of finding a maximal flow on G can be reduced to a maximal flow problem on a network with flow bounds on arcs only.

6.2. *Bilateral connections between nodes.* In many physical networks, a pair of nodes x_i and x_j may be joined by a 'bilateral' element, which can carry up to p units of flow in either direction, but which cannot carry flows in both directions simultaneously. (As an example, the transmission lines of an electric power system have this property.) How can one find a maximal flow between two nodes, in a network whose nodes are joined by elements of this kind?

6.3. There are p families $\alpha_1, \alpha_2, \ldots, \alpha_p$ which want to go for an excursion in q cars $\beta_1, \beta_2, \ldots, \beta_q$. Given the number s_i of members of each family α_i, and the number of seats d_j in each car β_j, is it possible to find a seating arrangement such that no two members of the same family are in the same car?

Formulate this problem as a maximal flow problem.

6.4. In a graph $G = (X, U)$, let A and B be two disjoint subsets of the node set X. Then (in accordance with the terminology of Sections 4.2

and 4.4) we say that a subset V of U is an (A, B)-*separating arc set* if every path from a node of A to a node of B traverses at least one arc in V. Similarly, a subset Y of $X - (A \cup B)$ is an (A, B)-*separating node set* if every path from a node of A to a node of B traverses at least one node in Y. Using a network flow method, find a separating arc set with the minimum number of arcs, and also a separating node set with the minimum number of nodes, for the node sets $A = \{x_2, x_6\}$, $B = \{x_4\}$ on the graph of Fig. 6.25.

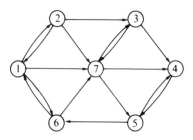

FIG. 6.25

6.5. The network of Fig. 6.26 represents a transportation system, with 'bilateral' connections between nodes (see Exercise 6.2); the first number associated with each connection is its capacity, the second its unit transportation cost. The nodes a and b represent producers of a commodity, capable of supplying up to 8 and 2 units respectively, at unit production costs of 5 and 6 respectively. The nodes d and e represent consumers, with demands of 6 and 1 units respectively. There is no supply or demand at node c.

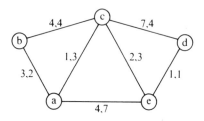

FIG. 6.26

Using the flow-augmentation method, determine the number of units to be supplied by each producer, and the amount to be sent along each branch of the network, in order to meet the demands at minimum total cost.

6.6. The network of Fig. 6.27(a) represents a transportation system, with bilateral connections between nodes. The number associated with each connection is its flow capacity; transportation costs can be neglected. At node a, a commodity is available in unlimited quantities; at nodes b, c, and d there is a demand for 3, 9, and 4 units respectively.

To meet the demands, it is necessary to increase the capacities of some connections in the network. The cost of increasing the capacities of the connections by one unit are indicated in Fig. 6.27(b). Determine a minimal-cost improvement of the network, such that all demands can be met.

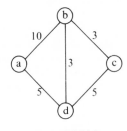

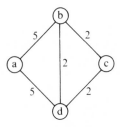

FIG. 6.27(a)   FIG. 6.27(b)

6.7. Let G be a network model of a transportation problem, as in Fig. 6.18, in which the sum of the supplies is greater than or equal to the sum of the demands:

$$\sum_{i=1}^{p} c_i \geq \sum_{i=1}^{q} d_i,$$

and let G' be the network derived from G by changing the cost of each arc (α_i, β_j) from λ_{ij} to

$$\lambda'_{ij} = \lambda_{ij} - \gamma_j \quad \text{where } \gamma_j = \min_{1 \leq i \leq p} \{\lambda_{ij}\}.$$

Prove that a flow **f** is a minimal-cost maximal flow of G if and only if **f** is a minimal-cost maximal flow of G', and determine the difference between the total costs of the minimal-cost maximal flows in the two networks.

(Note: the transformation of G to G' simplifies the transportation problem in that on G', at least one arc incident to each destination node β_j has a cost of zero; by sending the commodity along these arcs, one can easily obtain a zero-cost initial flow whose value is greater than zero. If the sum of the supplies is equal to the sum of the demands we can perform a further simplification, by changing the cost of each arc (α_i, β_j) from λ'_{ij} to

$$\lambda''_{ij} = \lambda'_{ij} - \delta_i \quad \text{where } \delta_i = \min_{1 \le j \le q} \{\lambda'_{ij}\}$$

in which case there will be at least one arc of zero cost incident from every source node α_i.)

6.8. Let **f** be a minimal-cost (but not necessarily maximal) flow on a network model G of a transportation problem (as in Fig. 6.18). We say that a source node α_i of G is *saturated* by **f** if the arc (s, α_i) is saturated; similarly, a destination node β_j is *saturated* by **f** if the arc (β_j, t) is saturated.

   Let $\tilde{G}(\mathbf{f})$ be the displacement network associated with **f**, and let $H(\mathbf{f})$ be the network obtained from $\tilde{G}(\mathbf{f})$ by deleting all the inverted arcs except those arcs (β_j, α_i) which are of non-zero capacity and whose endpoints are both saturated. Prove that if the network $\tilde{G}(\mathbf{f})$ has any flow-augmenting cycles, then at least one of its minimal-cost flow-augmenting cycles appears in $H(\mathbf{f})$.

6.9. There are three sources α_1, α_2, and α_3 of a commodity, which can supply 14, 6, and 9 units respectively. Three consumers β_1, β_2, and β_3 require 7, 10, and 8 units respectively. The relative costs of supplying one unit to each consumer, from each source, are given in the table below.

	β_1	β_2	β_3
α_1	5	12	3
α_2	9	5	5
α_3	4	6	3

   Using the flow-augmentation method, determine the number of units to be sent from each source to each consumer, in order to meet the demands at minimum total cost. (Employ the network simplification method of Exercise 6.7 to obtain an initial flow; to find flow-augmenting cycles use the simplified displacement networks, as defined in Exercise 6.8.)

6.10. A job in a workshop involves five tasks, to be assigned concurrently to five different machines. There are six machines available to perform these tasks; the time taken to set up each machine for a particular task is indicated in the table below. Find an assignment of machines to tasks for which the total set-up time is minimal.

	Tasks				
	1	2	3	4	5
1	5	4	8	4	5
2	2	4	3	2	3
3	4	1	7	3	4
4	8	3	1	4	3
5	9	4	9	6	7
6	3	8	7	6	4

Machines (rows 1–6)

6.11. (i) Transform the production planning problem of Example 6.5 into a minimal-cost maximal flow problem.

(ii) Find an optimal production pattern over four months, if the production costs and monthly demands are as follows.

Month	1	2	3	4
Demand	8	12	15	14
Unit production cost	£6000	£8000	£8000	£7000

The other production and storage capacities and costs are given below.

> Minimum monthly production: 10 units
> Maximum monthly production by overtime working: 5 units
> Additional cost of production by overtime working: £1000 per unit
> Storage capacity: 5 units
> Storage cost: £500 per unit per month.

(iii) In practice a 'handling cost' may be incurred in placing the commodity in the warehouse and removing it subsequently. Construct a new network model of the planning problem which takes account of handling costs.

6.12. A firm has two factories f_1 and f_2 which both manufacture the same commodity. Under normal working these factories produce 1000 and 500 units respectively, at unit production costs of £180 and £200 respectively. By overtime working the factory f_1 can produce up to 300 additional units per week, at a unit cost of £230, while the factory f_2 can produce up to 100 additional units, at a unit cost of £240.

The commodity is to be supplied to three customers c_1, c_2, and c_3, which require 700, 600, and 400 units respectively. The cost of sending one unit from each factory to each customer is given in the table below:

	c_1	c_2	c_3
f_1	£26	£20	£28
f_2	£32	£16	£18

Determine the number of units to be produced by overtime working at each factory, and the total number of units to be sent from each factory to each customer, in order to meet the consumer demands at minimum total cost.

6.13. The table below gives the departure times of passenger trains which run every day between stations A and B, and stations B and C. Each journey between A and B takes one hour, and each journey between B and C takes 90 minutes.

Origin	Destination	Departure time
A	B	0800, 0930, 1530 hours
B	A	1400, 1500, 1600 hours
B	C	1100, 1730 hours
C	B	1100, 1400 hours

Each train crew must finish its duties at the station from which it starts, not more than eight hours after its starting time. Find the minimum number of crews required, and a corresponding allocation of crews to trains. (Any number of crews can travel as passengers on a train.)

Additional notes and bibliography

The theory of network flows was largely developed by Ford and Fulkerson (1962). The subject is also treated in depth by M. Horps (see Roy 1970), Frank and Frisch (1971), and Lawler (1976c) who describe some extensions of the theory presented here, and further applications.

For efficient algorithms to solve maximal flow problems and discussions of the complexity of these algorithms see Dinic (1970), Edmonds and Karp (1972), Zadeh (1972), Hopcroft and Karp (1973) and Even and Tarjan (1975). The complexity of algorithms for finding minimal-cost maximal flows is also discussed by Zadeh (1973a, 1973b).

Algol programs based on the primal-dual method have been published, for the general minimal-cost maximal flow problem (Bray and Witzgall 1968), and also for the transportation problem in particular (Bayer 1966).

A method of obtaining good initial flows, in applying flow-augmentation methods to transportation problems, is described by Mueller-Merbach (1966).

For a linear programming approach to network-flow problems see Dantzig (1963) and Lawler (1976c). As an example to illustrate different methods of problem formulation and solution, the production planning problem of Example 6.5 has been solved by the Simplex method by Beale (1968), and Kaufmann (1967) solves a similar problem by dynamic programming.

With reference to Exercise 6.4, applications of network-flow methods to the determination of cut sets and related problems are discussed by Frank and Frisch (1971), Lawler (1973), and Colorni (1974).

Solutions to selected exercises

1.7. The operation ∘ is not idempotent, but it is commutative (since the operation table is symmetrical). The element ○ is a neutral element, and all elements are invertible, the inverses of ○, □ and △ being ○, △ and □ respectively. The set S does not contain a null element for ∘. The operation ∘ is cancellative (since in each row, and in each column, all the entries are distinct).

1.11. By setting $y = x \vee x$ in the second identity of L_4 we obtain

$$x \vee [x \wedge (x \vee x)] = x.$$

From the first identity of L4 it follows (by setting $y = x$) that in the above identity the expression in brackets is equal to x, hence $x \vee x = x$. The identity $x \wedge x = x$ can be proved by a similar argument, with ∨ and ∧ interchanged.

1.12. The conditions $w \leqslant x$ and $y \leqslant z$ can be expressed as $w \vee x = x$ and $y \vee z = z$ respectively. From these identities we obtain $(w \vee x) \vee (y \vee z) = x \vee z$, hence $(w \vee y) \vee (x \vee z) = x \vee z$, which can be written as $w \vee y \leqslant x \vee z$.

1.13. Since the two inequalities are dual, it suffices to prove the first. Now the relations $y \vee z \geqslant y$ and $y \vee z \geqslant z$ imply (by L11) that

$$x \wedge (y \vee z) \geqslant x \wedge y \quad \text{and} \quad x \wedge (y \vee z) \geqslant x \wedge z$$

and therefore (by L11, and the first part of L1)

$$x \wedge (y \vee z) \geqslant (x \wedge y) \vee (x \wedge z).$$

1.15. $x \leqslant z$ implies that $x \vee (y \wedge z) = (x \vee y) \wedge (x \vee z) = (x \vee y) \wedge z$.

1.16. $x \vee (\bar{x} \wedge y) = (x \vee \bar{x}) \wedge (x \vee y) = u \wedge (x \vee y) = x \vee y$;
$x \wedge (\bar{x} \vee y) = (x \wedge \bar{x}) \vee (x \wedge y) = \phi \vee (x \wedge y) = x \wedge y$.

2.7. (i) The tree search can be executed by repeated application of the two following rules.

Simplification rule: On a graph G, any node x_k which is the endpoint of a loop belongs to every feedback node set of G; these feedback node sets (with x_k removed) are the feedback node sets of the subgraph of G obtained by deleting x_k.

Specialization rule: Let x_k be any node of a graph G which is not the endpoint of a loop, and let us decompose the set M of feedback node sets of G into two subsets $M_k = \{Y \in M \mid x_k \in Y\}$ and $M_{\bar{k}} = \{Y \in M \mid x_k \notin Y\}$. The members of M_k, with x_k removed, are the feedback node sets of the subgraph G_k of G obtained by deleting x_k; while $M_{\bar{k}}$ is the set of feedback node sets of the graph $G_{\bar{k}}$ which is obtained from G by (a) joining each predecessor x_j of x_k to each successor x_l of x_k by an arc (x_j, x_l)—unless G contains this arc already—and then (b) removing the node x_k.

(ii) To find a feedback node set of minimum cardinality it is convenient first to apply the above simplification rule until all loops have been removed, and then to apply the following rule.

Second simplification rule: If G has a node x_k which is not the endpoint of a loop, and for which $\rho^+(x_k) \leq 1$ and $\rho^-(x_k) \leq 1$, then there exists a feedback node set of minimum cardinality which does not contain x_k (which implies that G can be replaced by the graph $G_{\bar{k}}$ defined above).

Note that the application of this rule may create loops, allowing further applications of the first simplification rule, which in turn may permit further applications of the second rule, and so on.

The search can also be made more efficient by applying the 'branch-and-bound' principle, in the following way. In the course of the search, we keep a record of the cardinality c of the smallest feedback node set yet discovered (initially, c is set to ∞); if for any sub-problem the total number of nodes already assigned to a feedback node set exceeds c then exploration of this sub-problem is terminated, since it could not yield any improved solutions.

3.1. Since (cf. (3.20)) $A^* = E \vee AA^*$,

$$\begin{bmatrix} B_{11} & B_{12} \\ B_{21} & B_{22} \end{bmatrix} = \begin{bmatrix} E_{11} & \Phi_{11} \\ \Phi_{12} & E_{22} \end{bmatrix} \vee \begin{bmatrix} A_{11} & A_{12} \\ A_{21} & A_{22} \end{bmatrix} \begin{bmatrix} B_{11} & B_{12} \\ B_{21} & B_{22} \end{bmatrix},$$

and therefore

$$B_{11} = E_{11} \vee A_{11}B_{11} \vee A_{12}B_{21}, \tag{1}$$

$$B_{12} = A_{11}B_{12} \vee A_{12}B_{22}, \tag{2}$$

$$B_{21} = A_{21}B_{11} \vee A_{22}B_{21}, \tag{3}$$

$$B_{22} = E_{22} \vee A_{21}B_{12} \vee A_{22}B_{22}. \tag{4}$$

The equations (1) and (2) can be rewritten as

$$B_{11} = A_{11}^*(A_{12}B_{21} \vee E_{11}), \tag{5}$$

$$B_{12} = A_{11}^*A_{12}B_{22}, \tag{6}$$

and substituting these expressions for B_{11} and B_{12} in (3) and (4) we obtain

$$B_{21} = (A_{21}A_{11}^*A_{12} \vee A_{22})B_{21} \vee A_{21}A_{11}^*, \tag{7}$$

$$B_{22} = E_{22} \vee (A_{21}A_{11}^*A_{12} \vee A_{22})B_{22}. \tag{8}$$

From (8),

$$B_{22} = (A_{21}A_{11}^*A_{12} \vee A_{22})^* \tag{9}$$

and from (7) and (8),

$$B_{21} = (A_{21}A_{11}^*A_{12} \vee A_{22})^*A_{21}A_{11}^* = B_{22}A_{21}A_{11}^*. \tag{10}$$

3.2. Let A be a stable matrix in $M_n(P)$, and let $A^* = [a_{ij}^*]$ denote its closure. By (3.23), $A^*A^* = A^*$ and therefore, by the definition of matrix multiplication,

$$a_{ii}^*a_{ii}^* \leqslant a_{ii}^*, \qquad (i = 1, 2, \ldots, n).$$

Since multiplication has the cancellative property it follows that

$$a_{ii}^* \leqslant e, \qquad (i = 1, 2, \ldots, n).$$

Now let γ be any elementary cycle on the graph of A, let k be the order of γ, and let x_i be any node on γ. Then by (3.44),

$$l(\gamma) \leqslant a_{ii}^k \leqslant a_{ii}^*.$$

It follows that $l(\gamma) \leqslant e$, as required.

3.3.

Activity	Start	a	b	c	d	e	f	g	h	i	j	Finish
Rank	0	1	5	1	2	4	5	2	3	6	3	7
Earliest starting time	0	0	51	0	5	35	51	5	5	54	15	65
Latest starting time	−5	21	46	−5	26	30	55	0	10	58	49	60
Slack time	−5	21	−5	−5	21	−5	4	−5	5	4	34	−5

The activity network has one critical path:

$$\text{Start} \to c \to g \to h \to e \to b \to \text{Finish},$$

whose length must be reduced by five time-units for the project to be completed on time.

3.4. The state diagram of the system is shown in Fig. E.1, in which the labels on the arcs are the costs of the corresponding transitions.

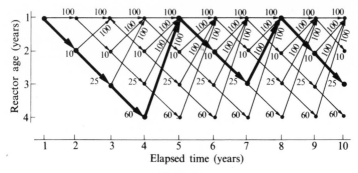

FIG. E.1

The adjacency matrix of this state diagram, regarded as a graph labelled with P_2, has off-diagonal blocks (cf. Example 3.22)

$$M^{(0)} = [100 \ \ 10], \qquad M^{(1)} = \begin{bmatrix} 100 & 10 & \infty \\ 100 & \infty & 25 \end{bmatrix},$$

$$M^{(2)} = \begin{bmatrix} 100 & 10 & \infty & \infty \\ 100 & \infty & 25 & \infty \\ 100 & \infty & \infty & 60 \end{bmatrix}$$

and

$$M^{(k)} = \begin{bmatrix} 100 & 10 & \infty & \infty \\ 100 & \infty & 25 & \infty \\ 100 & \infty & \infty & 60 \\ 100 & \infty & \infty & \infty \end{bmatrix} \quad \text{for } 3 \le k \le 9.$$

The successive $\mathbf{y}^{(k)}$-vectors obtained by the dynamic programming algorithm are listed below. (The arrows indicate, for each element $y_i^{(k)}$, those elements of the preceding vector $\mathbf{y}^{(k-1)}$ which determine its value.) In this particular problem there is only one optimal policy—

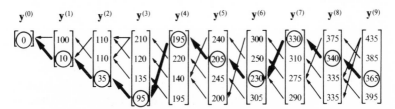

which is to replace the reactor at the end of the fourth and seventh years. The total cost under this policy is £365 000.

3.5. The required paths are obtained by solving the set of equations $\mathbf{y} = A'\mathbf{y} \vee \mathbf{e}_2$, where A is the adjacency matrix of the graph, using the path algebras P_2, P_5, and P_4 of Section 3.2. The $\mathbf{y}$-vectors and $\mathbf{s}$-vectors (as defined in Exercise 3.25) obtained in each case are

(i)
$$\mathbf{y} = \begin{bmatrix} 0.7 \\ 0 \\ 0.2 \\ 0.5 \\ 0.8 \\ 0.3 \end{bmatrix}, \quad \mathbf{s} = \begin{bmatrix} 4 \\ 0 \\ 2 \\ 2 \\ 1 \\ 3 \end{bmatrix};$$

(ii)
$$\mathbf{y} = \begin{bmatrix} 0.8 \\ \infty \\ 0.5 \\ 0.6 \\ 0.6 \\ 0.5 \end{bmatrix}, \quad \mathbf{s} = \begin{bmatrix} 2 \\ 0 \\ 4 \\ 1 \\ 4 \\ 5 \end{bmatrix};$$

(iii)
$$\mathbf{y} = \begin{bmatrix} 0.8 \\ 1 \\ 0.25 \\ 0.5 \\ 0.3 \\ 0.2 \end{bmatrix}, \quad \mathbf{s} = \begin{bmatrix} 2 \\ 0 \\ 4 \\ 2 \\ 4 \\ 4 \end{bmatrix}.$$

The $\mathbf{s}$-vectors give the sequences of nodes on the required paths *in reverse order*, since to solve the problem we have effectively reversed all the arcs of the graph.

3.6. The Hamiltonian cycles on an n-node graph G can be obtained by finding all the elementary cycles which terminate on some arbitrarily chosen node of G, and selecting from these all the cycles of order n. Using the path algebra P_8, the elementary cycles terminating on a node x_i are given by the ith entry of the vector $\mathbf{y}$ which satisfies the equation $\mathbf{y} = A\mathbf{y} \vee \mathbf{b}$, where A is the adjacency matrix of G and $\mathbf{b}$ is its ith column vector (see Table 3.2).

The successive $\mathbf{y}$-vectors obtained by the double-sweep method are listed below, for the case where $\mathbf{b}$ is the last column of A. (The brackets in the $\mathbf{y}$-vectors indicate the strings which are carried over from previous iterations.)

$$\mathbf{y}^{(0)} = \mathbf{b} \qquad \mathbf{y}^{(\frac{1}{2})} \qquad\qquad \mathbf{y}^{(1)} \qquad\qquad\qquad \mathbf{y}^{(1\frac{1}{2})}$$

$$\begin{bmatrix} b \\ \\ f \\ \\ \\ \end{bmatrix} \begin{bmatrix} (b), af \\ \\ (f) \\ \\ \\ \end{bmatrix} \begin{bmatrix} (b, af) \\ cb, caf \\ (f), dcb \\ gcb, gcaf, hf, hdcb \\ igcb, igcaf, ihf, ihdcb \end{bmatrix} \begin{bmatrix} (b, af), adcb \\ (cb, caf) \\ (f, dcb), egcb \\ (gcb, gcaf, hf, hdcb) \\ (igcb, igcaf, ihf, ihdcb) \end{bmatrix}$$

From the final entry in the $\mathbf{y}$-vector we find that the graph has two Hamiltonian cycles, igcaf, and ihdcb.

3.7. (ii) Since $E \vee L \leqslant L^* \leqslant A^*$ and $U \leqslant L^* U \leqslant A^*$ we have $E \vee L \vee U \leqslant A^*$, and therefore $\mathbf{y}_J^{(k)} \leqslant \mathbf{y}_{GS}^{(k)} \leqslant A^* \mathbf{b}$.

(iii) In case (a) we have $U = \phi$ and therefore

$$\mathbf{y}_J^{(k)} = (E \vee L)^k \mathbf{b} \quad \text{and} \quad \mathbf{y}_{GS}^{(k)} = \mathbf{y}_{ds}^{(k)} = L^* \mathbf{b};$$

the Gauss–Seidel and double-sweep methods become identical, and require only one iteration. In case (b), where $L = \phi$,

$$\mathbf{y}_J^{(k)} = \mathbf{y}_{GS}^{(k)} = (E \vee U)^k \mathbf{b} \quad \text{and} \quad \mathbf{y}_{ds}^{(k)} = U^* \mathbf{b};$$

the Jacobi and Gauss–Seidel methods become identical, and the double-sweep method requires only one iteration. In case (c) we have $L^* = E \vee L$ and $U^* = E \vee U$ and consequently

$$\mathbf{y}_J^{(k)} = (E \vee L \vee U)^k \mathbf{b} \quad \text{and} \quad \mathbf{y}_{GS}^{(k)} = \mathbf{y}_{ds}^{(k)} = (E \vee L \vee U \vee LU)^k \mathbf{b};$$

the Gauss–Seidel and double-sweep methods again become identical.

3.8. (i) $(w \vee xyz)^* = (w^* xyz)^* w^*$ (by (3.27))

$$= (e \vee w^* x (yzw^* x)^* yz) w^* \quad \text{(by (3.24))}$$

$$= w^* \vee w^* x (yzw^* x)^* yzw^*.$$

(ii) $B^* = (A \vee \mathbf{e}_i \sigma \mathbf{e}_j')^*$

$$= A^* \vee A^* \mathbf{e}_i (\sigma \mathbf{e}_j' A^* \mathbf{e}_i)^* \sigma \mathbf{e}_j' A^*$$

$$= A^* \vee \mathbf{c} (\sigma a_{ji}^*)^* \sigma \mathbf{d}.$$

3.10. The following algorithm will detect a negative cycle on G. (In this algorithm, the matrix $M = [m_{ij}]$ is initially a copy of the adjacency matrix of G.)

Step 1   [Initialize] $k \leftarrow 0$, $h \leftarrow 0$.

Step 2   [Augment k] $k \leftarrow k + 1$. If $k > n$ go to *End*.

Step 3   [Set row index] $i \leftarrow k$.

Step 4   [Augment row index] $i \leftarrow i + 1$. If $i > n$ return to *Step 2*.

Step 5   [Set column index] $j \leftarrow i$.

Step 6   [Test for negative cycle]. If $m_{ik} + m_{kj} < 0$ set $h \leftarrow i$ and go to *End*.

Step 7   [Augment column index] $j \leftarrow j + 1$. If $j > n$ return to *Step 4*.

Step 8   [Modify M]. Set $m_{ij} \leftarrow \min \{m_{ij}, m_{ik} + m_{kj}\}$ and $m_{ji} \leftarrow \min \{m_{ji}, m_{jk} + m_{ki}\}$. Return to *Step 7*.

End

If on termination $h = 0$ then G has no cycles of negative length. Otherwise, the subgraph H of G generated by $\{x_1, x_2, \ldots, x_k\} \cup \{x_h\}$ contains a cycle of negative length, and this cycle traverses x_h. To determine such a cycle, one can use Yen's method to find shortest paths from x_h to each of its predecessors, on the partial graph of H obtained by removing the arcs incident to x_h.

4.2. The weak closure matrix can be constructed using the simplified form (3.91) of the Jordan method. Here

$$A = \begin{bmatrix} \Omega & \{a\} & \Omega & \Omega & \\ \Omega & \Omega & \{b\} & \Omega & \Omega \\ \{c\} & \Omega & \Omega & \Omega & \Omega \\ \{d\} & \Omega & \{e\} & \Omega & \{f\} \\ \{g\} & \Omega & \Omega & \Omega & \Omega \end{bmatrix},$$

$$\hat{A} = \begin{bmatrix} \{a, b, c\} & \{a\} & \{a, b\} & \Omega & \Omega \\ \{b, c\} & \{a, b, c\} & \{b\} & \Omega & \Omega \\ \{c\} & \{a, c\} & \{a, b, c\} & \Omega & \Omega \\ \phi & \{a\} & \phi & \Omega & \{f\} \\ \{g\} & \{a, g\} & \{a, b, g\} & \Omega & \Omega \end{bmatrix}.$$

4.3. (iii) The rules for constructing $\tilde{T}$ from T are as follows. In case (a), remove from T the arc incident to x_i, and insert the arc (x_i, x_j). In case (b), $\tilde{T}$ is identical to T.
In case (c), find the node x_p which is of highest rank in the set $\hat{\Gamma}^-(x_i) \cap \hat{\Gamma}^-(x_j)$; then remove from T the arc incident to x_j, and insert the arc (x_p, x_j).

4.7. (i) G is its own leaf graph.
(ii) In case (a), $\tilde{G}_l$ is identical to G_l. In case (b), let X_r and X_s be the node sets of the leaves of G which contain the endpoints of e; then $\tilde{G}_l$ is the condensation of G_l which is obtained by coalescing all the nodes of G_l which lie on its elementary chain joining X_r to X_s.

5.1. Only four matched pairs can be obtained from the batch. As an example of a maximum matching we have $\{[2, 3], [4, 5], [6, 7], [9, 10]\}$.

5.4 Let G be the simple graph whose nodes correspond to the television stations, two nodes x_i and x_j being joined by an edge if the entry in the ith row and jth column of the table is a cross. Then a minimum coloration of G defines an appropriate assignment of frequencies (the different colours representing different frequencies). Using the simplification rules of Section 5.4.2 we find that four different frequencies are required, one possible assignment being as follows.

Station	1	2	3	4	5	6	7
Frequency	1	2	3	1	2	3	4

6.1. Split each node x_i (other than the source and sink) into two nodes x_i' and x_i'', in such a way that x_i' becomes the terminal endpoint of all arcs previously incident to x_i and x_i'' becomes the initial endpoint of all arcs previously incident from x_i, and join x_i' and x_i'' by an arc (x_i', x_i'') of capacity k_i.

6.2. Replace each bilateral connection by a pair of arcs with opposite orientations, both these arcs having the same capacity as the original connection. Then a flow of the required form can be obtained from any maximal flow on the modified network, by reducing the flows in the two arcs representing each bilateral connection by the lesser of their two values.

6.3. The network model is precisely the same as that for the shipping problem of Example 6.1, with each arc (α_i, β_j) having a capacity $w_{ij} = 1$. If a maximal flow does not saturate all the arcs incident from the source then the problem has no solution; otherwise, any maximal flow in which the arc flows are all integers defines a seating arrangement, through the rule that if the flow in arc (α_i, β_j) is non-zero then a member of family α_i is assigned to car β_j.

6.4. To obtain an (A, B)-separating arc set on a graph G, assign a unit capacity to each arc of G, and add to G

(i) a source node x_s, with arcs of infinite capacity from s to each node of A,

(ii) a sink node x_t, with arcs of infinite capacity from each node of B to x_t.

(iii) a return arc (x_t, x_s) of infinite capacity.

Let $\mathbf{f}$ be a maximal flow on this network, let X' be the set of all nodes which are accessible from x_s on the corresponding displacement network (with arcs of zero capacity removed), and let X'' be the set of remaining nodes. Then (X', X'') is an (A, B)-separating arc set of minimum cardinality.

Applying this technique to the graph of Fig. 6.25 we obtain the separating arc set $\{(x_3, x_4), (x_7, x_4), (x_7, x_5)\}$.

To obtain an (A, B)-separating node set assign an infinite capacity to each arc of G, assign a unit capacity to each node of G, and append to G a source and sink, as prescribed in (i)–(iii) above. Then replace each node of unit capacity by a pair of nodes joined by an arc of unit capacity (as in Exercise 6.1). Find a maximal flow on this network, and the corresponding cut (X', X''), as described above. This cut (in which every arc represents a node of unit capacity on the original graph) determines an (A, B)-separating node set of minimum cardinality.

Applying this technique to the graph of Fig. 6.25 we obtain the separating node set $\{x_3, x_7\}$.

6.5. A network G is constructed from the network of Fig. 6.26 as follows:

(i) each connection in Fig. 6.26 is replaced by a pair of arcs (as in Exercise 6.2);
(ii) a source node s is added, with arcs (s, a) and (s, b) having capacities 8 and 2 respectively and costs 5 and 6 respectively;
(iii) a sink node t is added, with arcs (d, t) and (e, t) having capacities 6 and 1 respectively and zero costs;
(iv) the source and sink are joined by a return arc (t, s) of infinite capacity and zero cost.

A minimal-cost maximal flow is then constructed on G. For this particular network, only one such flow exists; its non-zero arc flows are as follows.

Arc:	(s, a)	(s, b)	(a, b)	(a, c)	(a, e)	(b, c)	(c, d)	(e, d)	(d, t)	(e, t)
Flow:	5	2	2	1	2	4	5	1	6	1

The flows in the arcs (s, a) and (s, b) are the amounts to be supplied by the producers a and b respectively, and the flows in the arcs (d, t) and (e, t) are the amounts supplied to the consumers at d and e; it is evident that the consumer demands are met. The total cost of the flow is 95.

6.6. The problem involves finding a minimal-cost maximal flow through the network of Fig. E.2, where the first number on each connection represents its capacity, the second its cost; the flows in the connections of infinite capacity between the nodes a, b, c, and d then represent the required augmentations in the capacities of the corresponding existing connections.

A minimal-cost improvement (at a total cost of 7) is obtained by increasing the capacities of each of the connections a–d and b–c by one unit.

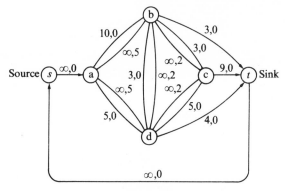

FIG. E.2

6.7. Let **f** be any maximal flow and let $\mathscr{L}$ and $\mathscr{L}'$ denote its costs on the networks G and G' respectively. Then

$$\mathscr{L}' = \sum_{i=1}^{p} \sum_{j=1}^{q} f_{ij}\lambda_{ij}' = \sum_{i=1}^{p} \sum_{j=1}^{q} f_{ij}(\lambda_{ij} - \gamma_j)$$

$$= \sum_{i=1}^{p} \sum_{j=1}^{q} f_{ij}\lambda_{ij} - \sum_{i=1}^{p} \sum_{j=1}^{q} f_{ij}\gamma_j.$$

Now

$$\sum_{i=1}^{p} \sum_{j=1}^{q} f_{ij}\lambda_j = \mathscr{L}$$

and

$$\sum_{i=1}^{p} \sum_{j=1}^{q} f_{ij}\gamma_j = \sum_{j=1}^{q} \left(\gamma_j \sum_{i=1}^{p} f_{ij} \right) = \sum_{j=1}^{q} \gamma_j d_j,$$

hence

$$\mathscr{L}' = \mathscr{L} - \sum_{j=1}^{q} \gamma_j d_j.$$

6.8. (i) Since flow-augmenting cycles are elementary, and traverse the arc (t, s), it is possible to remove from $\tilde{G}(\mathbf{f})$ all the inverted arcs incident to s, and all those which are incident from t, without destroying any flow-augmenting cycles.

(ii) A flow-augmenting cycle γ on $\tilde{G}(\mathbf{f})$ is composed of a path μ from s to t, together with the arc (t, s). The path μ is either a *direct path* of the form

$$(s, \alpha_i), (\alpha_i, \beta_j)(\beta_j, t)$$

or an *indirect* path of the form

$$(s, \alpha_{i_1}), (\alpha_{i_1}, \beta_{j_1}), (\beta_{j_1}, \alpha_{i_2}), \ldots, (\alpha_{i_r}, \beta_{j_r}), (\beta_{j_r}, t).$$

From the flow conservation condition it follows that if μ is indirect then, corresponding to each inverted arc $(\beta_{j_k}, \alpha_{i_{k+1}})$ on this path, $\tilde{G}(\mathbf{f})$ contains two inverted arcs $(\alpha_{i_{k+1}}, s)$ and (t, β_{j_k}) which are both of non-zero capacity. Since both these arcs have zero unit costs, and $\tilde{G}(\mathbf{f})$ has no cost-reducing cycles, all the segments of μ which are of the form

$$(s, \alpha_{i_1}), (\alpha_{i_1}, \beta_{j_1}), \ldots, (\beta_{j_k}, \alpha_{i_{k+1}}) \quad \text{where } 1 \le k < r,$$

and

$$(\beta_{j_k}, \alpha_{i_{k+1}}), \ldots, (\alpha_{i_r}, \beta_{j_r}), (\beta_{j_r}, t) \quad \text{where } 1 \le k < r,$$

are of non-negative cost. It follows that for every unsaturated source node $\alpha_{i_{k+1}}$ on μ, the path

$$(s, \alpha_{i_{k+1}}), (\alpha_{i_{k+1}}, \beta_{j_{k+1}}), \ldots, (\beta_{j_r}, t),$$

together with the arc (t, s), forms a flow-augmenting cycle whose cost is not greater than the cost of γ; similarly, for every unsaturated destination node β_{j_k} on μ, the path

$$(s, \alpha_{i_1}), (\alpha_{i_1}, \beta_{j_1}), \ldots, (\beta_{j_k}, t)$$

together with the arc (t, s) forms a flow-augmenting cycle, whose cost is not greater than the cost of γ. Hence all the inverted arcs of the form (β_j, α_i) whose endpoints are not both saturated can be removed from $\tilde{G}(\mathbf{f})$, without destroying all its minimal-cost paths.

6.9. Using the technique described in Exercise 6.7 we obtain the modified table of transportation costs shown in Fig. E.3. Then we arbitrarily

	β_1	β_2	β_3
α_1	1	7	0
α_2	5	0	2
α_3	0	1	0

	β_1	β_2	β_3
α_1	0	0	8
α_2	0	6	0
α_3	7	0	0

FIG. E.3   FIG. E.4

choose a feasible flow, using the zero-cost arcs, which is shown in Fig. E.4. Starting with this flow we then apply the flow-augmentation method. The displacement networks (simplified in the manner described in Exercise 6.8) and the successive flows are shown in Fig. E.5. (Only two flow augmentations are required.)

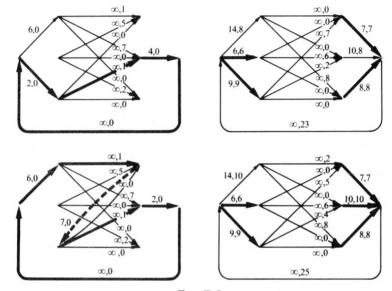

FIG. E.5

6.10. The solution of this problem can be simplified by the techniques of Exercises 6.7 and 6.8. In an optimal assignment, the total time spent setting up machines is 12. There are two such assignments:

Task:	1	2	3	4	5		Task:	1	2	3	4	5
Machine:	2	3	4	1	6		Machine:	6	3	4	2	1

6.11. The numbers of units to be produced by overtime working in the months 1, 2, 3, and 4 are 3, 0, 2, and 4, respectively.

6.12. The problem involves the determination of a minimal-cost circulation on the network of Fig. E.6, in which the three numbers on each arc are its lower flow bound, capacity and cost (all supplies and demands having been scaled down by a factor of 100 for convenience). By applying the transformation described in Sections 6.8.3 and 6.8.4, this problem can be reduced to a minimal-cost

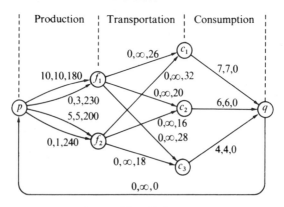

FIG. E.6

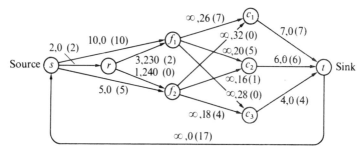

FIG. E.7

maximal flow problem on the network of Fig. E.7, in which the first number on each arc is its capacity and the second is its cost. (The flows in the arcs (r, f_1) and (r, f_2) represent the amounts produced by overtime working.) This particular problem has a unique solution, for which the arc flows are indicated on Fig. E.7 in brackets.

The factory f_1 produces 200 units by overtime working, and the amount sent from each factory to each consumer is as follows:

	c_1	c_2	c_3
f_1	700	500	—
f_2	—	100	400

The total cost of production and transportation is £363 000.

6.13. The daily activity of a train crew can be visualized as the traversal of an elementary cycle on Fig. E.8, where

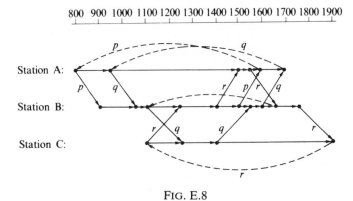

—Time———►

800 900 1000 1100 1200 1300 1400 1500 1600 1700 1800 1900

Station A:

Station B:

Station C:

FIG. E.8

(i) traversal of a horizontal arc drawn as a solid line represents time spent waiting in a station,

(ii) traversal of a diagonal arc represents a journey on a train, and

(iii) traversal of an arc drawn as a broken line represents time spent off duty.

Lower flow bounds, capacities, and costs are assigned to the arcs as follows:

(i) each horizontal arc is assigned a lower flow bound of zero, infinite capacity and zero cost;

(ii) each diagonal arc is assigned a lower flow bound of one, infinite capacity and zero cost;

(iii) each arc drawn as a broken line is assigned a lower flow bound of zero, infinite capacity and a cost of one.

Any decomposition into elementary cyclic flows of an integer-valued minimal-cost circulation on this network determines an appropriate allocation of crews to trains. In this problem, three crews are required; a feasible allocation of three crews p, q, and r to the trains is indicated on the diagram.

Bibliography

AHO, A. V. and ULLMAN, J. D. (1977). *Principles of Compiler Design.* Addison-Wesley, Reading, Mass.

—— HOPCROFT, J. E. and ULLMAN, J. D. (1974). *The design and analysis of computer algorithms.* Addison-Wesley, Reading, Mass.

AKKOYUNLU, E. A. (1973). The enumeration of maximal cliques of large graphs. *SIAM J. Comput.* **2**, 1–6.

AUGUSTSON, J. G. and MINKER, J. (1970). An analysis of some graph theoretical cluster techniques. *J. Ass. Comput. Mach.* **17**, 571–88. (Correction: *J. Ass. Comput. Mach.* **19** (1972), 244–7.)

BACKHOUSE, R. C. and CARRÉ, B. A. (1975). Regular algebra applied to path-finding problems. *J. Inst. Math. & its Appl.* **15**, 161–86.

BALINSKI, M. L. (1969). Labelling to obtain a maximum matching. In *Combinatorial mathematics and its applications* (edited by R. C. Bose and T. A. Dowling), 585–602. University of North Carolina Press, Chapel Hill.

BAYER, G. (1966). Algorithm 293: Transportation problem. *Communs Ass. Comput. Mach.* **9**, 869–71; **10**, 453; **11**, 271–2.

BEALE, E. M. L. (1968). *Mathematical programming in practice.* Pitman, London.

BELLMAN, R. E. (1957). *Dynamic programming.* Princeton University Press, N.J.

—— (1958). On a routing problem. *Q. Appl. Math.* **16**, 87–90.

BELLMORE, M. and HONG, S. (1974). Transformation of multisalesman problem to the standard traveling salesman problem. *J. Ass. Comput. Mach.* **21**, 500–4.

—— and NEMHAUSER, G. L. (1968). The traveling salesman problem: a survey. *Ops Res.* **16**, 538–58.

BENZAKEN, C. (1968). Structures algébriques des cheminements: pseudotreillis, gerbiers de carré nul. In *Network and switching theory* (edited by G. Biorci), 40–57. Academic Press, London.

BERGE, C. (1957). Two theorems in graph theory. *Proc. Natn. Acad. Sci. U.S.A.* **43**, 842–4.

—— (1976). *Graphs and hypergraphs* (2nd edn.). North-Holland, Amsterdam.

BIGGS, N. L., LLOYD, E. K., and WILSON, R. J. (1976). *Graph theory 1736–1936.* Clarendon Press, Oxford.

BIRKHOFF, G. (1967). *Lattice theory* (3rd edn). American Mathematical Society, Providence, R.I.

—— and MAC LANE, S. (1977). *A survey of modern algebra* (4th edn). Macmillan, New York.

BITNER, J. R. and REINGOLD, E. M. (1975). Backtrack programming techniques. *Communs Ass. Comput. Mach.* **18**, 651–6.

BRAY, T. A. and WITZGALL, C. (1968). Algorithm 336: Netflow. *Communs Ass. Comput. Mach.* **11**, 631–2; **13**, 192.

BRON, C. and KERBOSCH, J. (1973). Algorithm 457: Finding all cliques of an undirected graph. *Communs Ass. Comput. Mach.* **16**, 575–7.

BRUCKER, P. (1974). *Theory of matrix algorithms.* Mathematical systems in economics, Vol. 13. Verlag Anton Hain, Meisenheim am Glan.

CARRÉ, B. A. (1969). A matrix factorization method for finding optimal paths through networks. *I.E.E. Conference Publication No. 51 (Computer-aided design)*, 388–97.

—— (1970). An elimination method for minimal-cost network flow problems. In *Large sparse sets of linear equations* (edited by J. K. Reid), 191–209. Academic Press, London.

—— (1971). An algebra for network routing problems. *J. Inst. Math. & its Appl.* **7**, 273–94.

CHEN, W.-K. (1971). *Applied graph theory.* North-Holland, Amsterdam.

CHERITON, D. and TARJAN, R. E. (1976). Finding minimum spanning trees. *SIAM J. Comput.* **5**, 704–14

CHRISTOFIDES, N. (1971). An algorithm for the chromatic number of a graph. *Comput. J.* **14**, 38–9.

CLARE, C. (1973). *Designing logic systems using state machines*, McGraw-Hill, New York.

COLORNI, A. (1974). An algorithm for the determination of an optimal cutset in a graph. *Ric. Autom.* **5**, 41–51.

CORNEIL, D. G. (1971). An n^2 algorithm for determining the bridges of a graph. *Inf. Process. Lett.* **1**, 51–5.

—— (1974). The analysis of graph-theoretical algorithms. *Proc. Fifth South-Eastern Conference on Combinatorics, Graph Theory and Computing*, 3–38. Utilitas Mathematica Publishing Inc., Winnipeg.

—— and GRAHAM, B. (1973). An algorithm for determining the chromatic number of a graph. *SIAM J. Comput.* **2**, 311–18.

CRUON, R. and HERVÉ, P. (1965). Quelques résultats relatifs à une structure algébrique et à son application au problème central d'ordonnancement. *Revue Française de Recherche Opérationnelle* No. 34, 3–19.

CUNINGHAME-GREEN, R. A. (1962). Describing industrial processes with interference and approximating their steady-state behaviour. *Opl Res. Q.* **13**, 95–100.

—— (1976). Projections in minimax algebra. *Math. Programming* **10**, 111–23.

DANTZIG, G. B. (1963). *Linear programming and extensions.* Princeton University Press, Princeton, N.J.

—— (1966). All shortest routes in a graph. In *Theory of graphs (International Symposium, Rome 1966)*, 91–2. Gordon and Breach, New York.

DERNIAME, J.-C. and PAIR, C. (1971). *Problèmes de cheminement dans les graphes.* Dunod, Paris.

DIJKSTRA, E. W. (1959). A note on two problems in connection with graphs. *Num. Math.* **1**, 269–71.

DINIC, E. A. (1970). Algorithm for solution of a problem of maximum flow in a network with power estimation. *Soviet Math. Dokl.* **11**, 1277–80.

DUBREIL-JACOTIN, M. L., LESIEUR, L., and CROISOT, R. (1953). *Leçons sur la théorie des treillis, des structures algébriques ordonnées et des treillis géométriques.* Gauthier-Villars, Paris.

DUFF, I. S. (1977). A survey of sparse matrix research. *Proc. IEEE* **65**, 500–35.

DUFFIN, R. J. (1959). An analysis of the Wang algebra of networks. *Trans. Am. Math. Soc.* October, 114–30.

EDMONDS, J. (1965). Paths, trees and flowers. *Can. J. Math.* **17**, 449–67.

—— and KARP, R. M. (1972). Theoretical improvements in algorithmic efficiency for network flow problems. *J. Ass. Comput. Mach.* **19**, 248–64.

ELMAGHRABY, S. E. (1970). *Some network models in management science.* Lecture notes in operations research and mathematical systems, Vol. 29. Springer-Verlag, Berlin.

EVEN, S. and KARIV, O. (1975). An $O(n^{2.5})$ algorithm for maximum matching in general graphs. *Proc. 16th Annual Symp. on Foundations of Computer Science*, 100–12. IEEE, New York.

—— and TARJAN, R. E. (1975). Network flow and testing graph connectivity. *SIAM J. Comput.* **4**, 507–518.

FADDEEVA, V. N. (1959). *Computational methods of linear algebra.* Dover, New York.

FILLMORE, J. P. and WILLIAMSON, S. G. (1974). On backtracking: a combinatorial description of the algorithm. *SIAM J. Comput.* **3**, 41–55.

FISHER, A. C., LIEBMAN, J. S., and NEMHAUSER, G. L. (1968). Computer construction of project networks. *Communs Ass. Comput. Mach.* **11**, 493–7.

FLOYD, R. W. (1962). Algorithm 97: shortest path. *Communs Ass. Comput. Mach.* **5**, 345.

—— (1967). Nondeterministic algorithms. *J. Ass. Comput. Mach.* **14**, 636–44.

FONTAN, G. (1974). Sur les performances d'algorithmes de recherche de chemins minimaux dans les graphes clairsemés. *Revue Française d'Automatique, Informatique et Recherche Opérationnelle* **8**, V–2, 31–7.

FORD, L. R. and FULKERSON, D. R. (1962). *Flows in networks.* Princeton University Press, Princeton, N.J.

FOX, L. (1964). *An introduction to numerical linear algebra.* Oxford University Press.

FRANK, H. and FRISCH, I. T. (1971). *Communication, transmission and transportation networks.* Addison-Wesley, Reading, Mass.

FRATTA, L. and MONTANARI, U. (1975). A vertex elimination algorithm for enumerating all simple paths in a graph. *Networks* **5**, 151–77.

GABOW, H. N. (1976). An efficient implementation of Edmonds' algorithm for maximum matching on graphs. *J. Ass. Comput. Mach.* **23**, 221–34.

—— (1977). Two algorithms for generating weighted spanning trees in order. *SIAM J. Comput.* **6**, 139–50.

GAREY, M. R. and JOHNSON, D. S. (1976). The complexity of near-optimal graph coloring. *J. Ass. Comput. Mach.* **23**, 43-9.

GARFINKEL, R. S. and NEMHAUSER, G. L. (1972). *Integer programming.* Wiley, New York.

GOLDEN, B. (1976). Shortest path algorithms: a comparison. *Ops Res.* **24**, 1164–8.

GOLOMB, S. W. and BAUMERT, L. D. (1965). Backtrack programming. *J. Ass. Comput. Mach.* **12**, 516–24.

GONDRAN, M. (1975). Algèbre linéaire et cheminement dans un graphe. *Revue Française d'Automatique, Informatique et Recherche Opérationnelle* **9**, V–1, 77–99.

GUARDABASSI, G. (1971). A note on minimal essential sets. *IEEE Trans. on Circuit Theory* **CT-18**, 557–60.

HAMMER, P. L. and RUDEANU, S. (1968). *Boolean methods in operations research.* Springer-Verlag, Berlin.

HARARY, F., NORMAN, R. Z., and CARTWRIGHT, D. (1965). *Structural models: an introduction to the theory of directed graphs.* John Wiley, New York.

HATCH, R. S. (1975). Bench marks comparing transportation codes based on primal simplex and primal-dual algorithms. *Ops Res.* **23**, 1167–72.

HECHT, M. S. (1977). *Flow analysis of computer programs.* Elsevier-North-Holland, New York.

HEDETNIEMI, S. T. (1971). Review No. 22063, *Comput. Rev.* **12**, 446–7.

HELD, M. and KARP, R. M. (1970). The traveling-salesman problem and minimum spanning trees. *Ops Res.* **18**, 1138–62.

—— —— (1971). The traveling-salesman problem and minimum spanning trees: Part II. *Math. Program.* **1**, 6–25.

HOFFMAN, A. J. and WINOGRAD, S. (1972). Finding all shortest distances in a directed network. *IBM J. Res. Develop.* **16**, 412–4.

HOPCROFT, J. E. and KARP, R. M. (1973). An $n^{5/2}$ algorithm for maximum matchings in bipartite graphs. *SIAM J. Comput.* **2**, 225–31.

—— and TARJAN, R. E. (1973). Efficient algorithms for graph manipulation. *Communs Ass. Comput. Mach.* **16**, 372–8.

HOUSEHOLDER, A. S. (1953). *Principles of numerical analysis.* McGraw-Hill, New York.

HU, T. C. and TORRES, W. T. (1969). Shortcut in the decomposition algorithm for shortest paths in a network. *IBM J. Res. Develop.* **13**, 387–90.

HULME, B. L. (1975). A lattice algebra for finding simple paths and cuts in a graph. *Proc. Sixth South-Eastern Conf. on Combinatorics, Graph Theory and Computing*, 419–28. Utilitas Mathematica Publishing Inc., Winnipeg.

JENSEN, P. A. and BELLMORE, M. (1969). An algorithm to determine the reliability of a complex system. *IEEE Trans. on Reliability* **R–18**, 169–74.

JOHNSON, D. B. (1975). Finding all the elementary circuits of a directed graph. *SIAM J. Comput.* **4**, 77–84.

JOHNSON, D. S. (1974). Worst-case behaviour of graph coloring algorithms. *Proc. Fifth South-Eastern Conference on Combinatorics, Graph Theory and Computing*, 513–28. Utilitas Mathematics Publishing Inc., Winnipeg.

JOHNSON, E. L. (1972). On shortest paths and sorting. *Proc. Ass. Comput. Mach. 25th Annual Conference, Boston, 1972*, Vol. 1, 510–17.

JOHNSTON, H. C. (1976). Cliques of a graph: variations on the Bron–Kerbosch algorithm. *Int. J. Comput. & Inf. Sci.* **5**, 209–38.

KARP, R. M. (1972). Reducibility among combinatorial problems. In *Complexity of computer computations* (edited by R. E. Miller and J. W. Thatcher), 85–103. Plenum Press, New York.

—— (1975*a*). On the computational complexity of combinatorial problems. *Networks* **5**, 45–68.

—— (1975*b*). The fast approximate solution of hard combinatorial problems. *Proc. Sixth South-Eastern Conference on Combinatorics, Graph Theory and Computing*, 15–31. Utilitas Mathematica Publishing Inc., Winnipeg.

KAUFMANN, A. (1967). *Graphs, dynamic programming and finite games.* Academic Press, New York.

—— and PICHAT, E. (1977). *Méthodes mathématiques non-numériques et leurs algorithmes.* Tome 1: *Algorithmes de recherche des éléments maximaux.* Tome 2: *Algorithmes de recherche de chemins et problèmes associés.* Masson, Paris.

KERSHENBAUM, A. and VAN SLYKE, R. (1972). Computing minimum spanning trees efficiently. *Proc. Ass. Comput. Mach. Conf.* (*1972*), 518–27.

KLEIN, M. (1967). A primal method for minimal cost flows. *Manage. Sci.* **14**, 205–20.

KNUTH, D. E. (1968). *The art of computer programming*, Vol. 1: *Fundamental algorithms.* Addison-Wesley, Reading, Mass.

——(1973). *The art of computer programming*, Vol. 3: *Sorting and searching.* Addison-Wesley, Reading, Mass.

——(1975). Estimating the efficiency of back-track programs. *Maths Comput.* **29**, 121–36.

——(1977). A generalisation of Dijkstra's algorithm. *Inf. Process. Lett.* **6**, 1–5.

KROFT, D. (1967). All paths through a maze. *Proc. IEEE* **55**, 88–90.

KRUSKAL, J. B. (1956). On the shortest spanning subtree of a graph and the traveling salesman problem. *Proc. Am. Math. Soc.* **7**, 48–50.

LAND, A. H. and STAIRS, S. W. (1967). The extension of the cascade algorithm to large graphs. *Manage. Sci.* **14**, 29–33.

LAVROV, S. S. (1961). Store economy in closed operator schemes. *J. USSR, Comput. Maths. and Math. Physics* (English Translations) **1**, 810–28.

LAWLER, E. L. (1973). Cutsets and partitions of hypergraphs. *Networks* **3**, 275–85.

——(1976*a*). Introduction to the complexity of algorithms. Chapter 2 of *Applied computation theory: analysis, design, modelling* (edited by R. T. Yeh). Prentice-Hall, Englewood Cliffs, N.J.

——(1976*b*). A note on the complexity of the chromatic number problem. *Inf. Process. Lett.* **5**, 66–7.

——(1976*c*). *Combinatorial optimization: networks and matroids.* Holt, Rinehart, and Winston, New York.

—— and WOOD, D. E. (1966). Branch-and-bound methods: a survey. *Ops Res.* **14**, 699–719.

LEAVENWORTH, B. (1961). Algorithm 40: Critical path scheduling. *Communs Ass. Comput. Mach.* **4**, 152; **4**, 392; **5**, 513; **7**, 349.

LEE, C. Y. (1961). An algorithm for path connections and its applications. *IRE Trans.* **EC–10**, 346–65.

LITTLE, J. D. C., MURTY, K. G., SWEENEY, D. W., and KAREL, C. (1963). An algorithm for the traveling-salesman problem. *Ops Res.* **11**, 972–89.

LOGRIPPO, L. (1972). Renaming in program schemas. *IEEE Conference Record 13th Annual Symposium on Switching and Automata Theory*, 67–70.

——(1978). Renamings and economy of memory in program schemata. *J. Ass. Comput. Mach.* **25**, 10–22.

LOWRY, E. S. and MEDLOCK, C. W. (1969). Object code optimization, *Communs Ass. Comput. Mach.* **12**, 13–22.

LUNTS, A. G. (1950). The application of Boolean matrix algebra to the analysis and synthesis of relay-contact networks. (In Russian.) *Dokl. Akad. Nauk. SSSR* **70**, 421–3.

MCILROY, M. D. (1969). Algorithm 354: Generator of spanning trees. *Communs Ass. Comput. Mach.* **12**, 511.

MCNAUGHTON, R. and YAMADA, H. (1960). Regular expressions and state graphs for automata. *IRE Trans. on Electronic Computers* **EC-9**, 39–47.

MARTELLI, A. (1974). An application of Regular Algebra to the enumeration of the cut sets of a graph. *Information Processing 74*, 511–5. North-Holland, Amsterdam.

——(1976). A Gaussian elimination algorithm for the enumeration of cut sets in a graph *J. Ass. Comput. Mach.* **23**, 58–73.

MATULA, D. W., MARBLE, G., and ISAACSON, J. D. (1972). Graph colouring algorithms. In *Graph theory and computing* (edited by R. C. Read), 109–22. Academic Press, London.

MINIEKA, E. and SHIER, D. R. (1973). A note on an algebra for the k best routes in a network. *J. Inst. Math. & its Appl.* **11**, 145–9.

MINOUX, M. (1976). Structures algébriques généralisées des problèmes de cheminement dans les graphes. *Revue Française d'Automatique, Informatique et Recherche Opérationnelle* **10**, 33–62.

MINTY, G. J. (1965). A simple algorithm for listing all the trees of a graph. *IEEE Trans. on Circuit Theory* **CT-12**, 120.

MITCHEM, J. (1976). On various algorithms for estimating the chromatic number of a graph. *Comput. J.* **19**, 182–3.

MITTEN, L. G. (1970). Branch-and-bound methods: general formulation and properties. *Ops Res.* **18**, 24–34.

MOISIL, GR. C. (1960). Asupra unor representări ale grafurilor ce intervin în probleme de economia transporturilor. *Comunle Acad. Rep. Pop. Rom.* **10**, 647–52.

MONTALBANO, M. (1967). High-speed calculation of the critical paths of large networks. *IBM Systems J.* **6**, 163–91.

MOON, J. W. and MOSER, L. (1965). On cliques in graphs. *Israel J. Math.* **3**, 123–8.

MOYLES, D. M. and THOMPSON, G. L. (1969). An algorithm for finding a minimum equivalent graph of a digraph. *J. Ass. Comput. Mach.* **16**, 455–60.

MUELLER-MERBACH, H. (1966). An improved starting algorithm for the Ford–Fulkerson approach to the transportation problem. *Manage. Sci.* **13**, 97–104.

MULLIGAN, G. D. and CORNEIL, D. G. (1972). Corrections to Bierstone's algorithm for generating cliques. *J. Ass. Comput. Mach.* **19**, 244–7.

MUNRO, I. (1971). Efficient determination of the transitive closure of a directed graph. *Inf. Process. Lett.* **1**, 56–8.

MURCHLAND, J. D. (1965). A new method for finding all elementary paths in a complete directed graph. *Report LSE-TNT-22, London School of Economics.*

—— (1967). The effect of increasing or decreasing the length of a single arc on all shortest distances in a graph. *Report LBS-TNT-26, London School of Economics.*

ORE, O. (1962). *Theory of graphs (Amer. Math. Soc. Colloquium Publications, Vol. 38)*, American Mathematical Society, Providence, R.I.

—— (1967). *The four-color theorem.* Academic Press, New York.

OSTEEN, R. E. (1974). Clique detection algorithms based on line addition and line removal. *SIAM J. Appl. Math.* **26**, 126–35.

—— and TOU, J. T. (1973). A clique-detection algorithm based on neighbourhoods in graphs. *Int. J. Comput. and Inf. Sci.* **2**, 257–68.

PATON, K. (1971). An algorithm for the blocks and cutnodes of a graph. *Communs Ass. Comput. Mach.* **14**, 468–75, Corrigendum *Communs Ass. Comput. Mach.* **14**, 592.

PAULL, M. C. and UNGER, S. H. (1959). Minimizing the number of states in incompletely specified sequential switching functions. *IRE Trans. on Electronic Computers* **EC-8**, 356–67.

PECK, J. E. L. and WILLIAMS, M. R. (1966). Algorithm 286: Examination scheduling. *Communs Ass. Comput. Mach.* **9**, 433–4.

PETEANU, V. (1967). An algebra of the optimal path in networks. *Mathematica (Cluj)* **9**, 335–42.

—— (1969). Optimal paths in networks and generalizations (I). *Mathematica (Cluj)* **11**, 311–27.

—— (1970). Optimal paths in networks and generalizations (II). *Mathematica (Cluj)* **12**, 159–86.

PIERCE, A. R. (1975). Bibliography on algorithms for shortest path, shortest spanning tree, and related circuit routing problems (1956–1974). *Networks* **5**, 129–49.

POLYA, G. (1957). *How to solve it: a new aspect of mathematical method* (2nd ed). Princeton University Press, Princeton, N.J.,

PRIM, R. C. (1957). Shortest connection networks and some generalizations. *Bell Syst. Tech. J.* **36**, 1389–1401.

PURDOM, P. W. and MOORE, E. F. (1972). Algorithm 430: Immediate predominators in a directed graph. *Communs Ass. Comput. Mach.* **15**, 777–8.

READ, R. C. and TARJAN, R. E. (1975). Bounds on backtrack algorithms for listing cycles, paths, and spanning trees. *Networks* **5**, 237–52.

ROBERT, P. (1971). An algorithm for finding the essential sets of arcs of certain graphs. *J. Combinatorial Theory (Series B)* **10**, 288–98.

—— and FERLAND, J. (1968). Généralisation de l'algorithme de Warshall. *Revue Française d'Informatique et de Recherche Opérationnelle* **2**, 71–85.

ROBERTS, S. M. and FLORES, B. (1966). Systematic generation of Hamiltonian circuits. *Communs Ass. Comput. Mach.* **9**, 690–4.

RODIONOV, V. V. (1968). The parametric problem of shortest distances. *USSR Comput. Math. & Math. Phys.* **8**, 336–43.

ROSCHKE, S. I. and FURTADO, A. L. (1973). An algorithm for obtaining the chromatic number and an optimal coloring of a graph. *Inf. Process. Lett.* **2**, 34–8.

ROSE, D. J. (1972). A graph-theoretic study of the numerical solution of sparse positive definite systems of linear equations. In *Graph theory and computing* (edited by R. C. Read), 183–217. Academic Press, London.

ROY, B. (1959). Transitivité et connexité. *C.R. Acad. Sci. Paris* **249**, 216–8.

—— (1969, 1970). *Algèbre moderne et théorie des graphes*, Vol. 1 (1969) and Vol. 2 (1970). Dunod, Paris. Vol. 1 has been translated (1978): *Modern algebra and graph theory applied to management.* Springer-Verlag.

—— (1975). Chemins et circuits: énumération et optimisation. In *Combinatorial programming: methods and applications* (edited by B. Roy). D. Reidel Publishing Company, Dordrecht.

RUBIN, F. (1974). A search procedure for Hamiltonian paths and circuits. *J. Ass. Comput. Mach.* **21**, 576–80.

SCHAEFER, M. (1973). *A mathematical theory of global program optimisation.* Prentice-Hall, Englewood Cliffs, N. J.

SEPPANEN, J. J. (1970). Algorithm 399: Spanning tree. *Communs Ass. Comput. Mach.* **13**, 621–2.

SHIER, D. R. (1976). Iterative methods for determining the *k* shortest paths in a network. *Networks* **6**, 205–29.

SMITH, G. W. and WALFORD, R. B. (1975). The identification of a minimal feedback vertex set of a directed graph. *IEEE Trans. on Circuits and Systems* **CAS–22**, 9–14.

SZWARCFITER, J. L. and LAUER, P. E. (1976). A search strategy for the elementary cycles of a directed graph. *BIT* **16**, 192–204.

TABOURIER, Y. (1972). Un algorithme pour le problème d'affectation. *Revue Française d'Automatique, Informatique et Recherche Opération-nelle* **6**, V–3, 3–16.

TARJAN, R. E. (1972). Depth-first search and linear graph algorithms. *SIAM J. Comput.* **1**, 146–60.

—— (1973). Enumeration of the elementary circuits of a directed graph. *SIAM J. Comput.* **2**, 211–6.

— (1974*a*). Finding dominators in directed graphs. *SIAM J. Comput.* **3**, 62–89.

—— (1974*b*). A note on finding the bridges of a graph. *Inf. Process. Lett.* **2**, 160–1.

——and TROJANOWSKI, A. E. (1977). Finding a maximum independent set. *SIAM J. Comput.* **8**, 537–46.

TARRY, M. G. (1895). Le problème des labyrinthes. *Nouv. Annls Math.* **14**, 187–90.

TEWARSON, R. P. (1973). *Sparse matrices.* Academic Press, New York.

TIERNAN, J. C. (1970). An efficient search algorithm to find the elementary circuits of a graph. *Communs Ass. Comput. Mach.* **13**, 722–6.

TOMESCU, I. (1966). Sur les méthodes matricielles dans la théorie des réseaux. *C.R. Acad. Sci. Paris* **263**, Série A, 826–9.

—— (1968). Sur l'algorithme matriciel de B. Roy. *Revue Française d'Informatique et de Recherche Opérationnelle* **2**, 87–91.

TOMIZAWA, N. (1971). On some techniques useful for solution of transportation network problems. *Networks* **1**, 173–94.

TRENT, H. M. (1954). A note on the enumeration and listing of all possible trees in a connected linear graph. *Proc. Natn. Acad. Sci. U.S.A.* **40**, 1004–7.

TSUKIYAMA, S., IDE, M., ARIYOSHI, H., and SHIRAKAWA, I. (1977). A new algorithm for generating all the maximal independent sets. *SIAM J. Comput.* **6**, 505–17.

—— SHIRAKAWA, I., and OZAKI, H.(1975). An algorithm for generating the cycles of a digraph. *Electron. & Commun. Japan* **58–A**, 8–15.

VARGA, R. S. (1962). *Matrix iterative analysis.* Prentice-Hall, Englewood Cliffs.

WALKER, R. J. (1960). An enumerative technique for a class of combinatorial problems. *Proc. Sympos. Appl. Math. Vol. 10* (*Combinatorial analysis*), 91–4. American Mathematical Society, Providence, R.I.

WANG, C. C. (1974). An algorithm for the chromatic number of a graph. *J. Ass. Comput. Mach.* **21**, 385–91.

WARSHALL, S. (1962). A theorem on Boolean matrices. *J. Ass. Comput. Mach.* **9**, 11–3.

WELLS, M. B. (1971). *Elements of combinatorial computing.* Pergamon, Oxford.

WELSH, D. J. A. and POWELL, M. B. (1967). An upper bound for the chromatic number of a graph and its application to timetabling problems. *Comput. J.* **10**, 85–6.

WHITE, D. J. (1969). *Dynamic programming.* Oliver and Boyd, Edinburgh.

WHITNEY, V. K. M. (1972). Algorithm 422: Minimal spanning tree. *Communs Ass. Comput. Mach.* **15**, 273–4.

WILLIAMS, M. R. (1970). The colouring of very large graphs. In *Combinatorial structures and their applications* (edited by R. Guy, H. Hanani, N. Sauer, and J. Schonheim), 477–8. Gordon and Breach, New York.

WITZGALL, C. and ZAHN, C. T. (1965). Modification of Edmonds' maximum matching algorithm. *J. Res. NBS* **69B**, 91–8.

WONGSEELASHOTE, A. (1976). An algebra for determining all path-values in a network, with application to k-shortest path problems. *Networks* **6**, 307–34.

WOOD, D. C. (1969). A technique for colouring a graph applicable to large scale timetabling problems. *Comput. J.* **12**, 317–9.

YAO, A. C. (1975). An $O(|E| \log \log |V|)$ algorithm for finding minimum spanning trees. *Inf. Process. Lett.* **4**, 21–3.

YEN, J. Y. (1970). An algorithm for finding shortest routes from all source nodes to a given destination in general networks. *Q. Appl. Math.* **27**, 526–30.

—— (1975). *Shortest path network problems.* Mathematical systems in economics, Vol. 18, Verlag Anton Hain, Meisenheim am Glan.

YOELI, M. (1961). A note on a generalization of Boolean matrix theory. *Am. Math. Mon.* **68**, 552–7.

ZADEH, N. (1972). Theoretical efficiency of the Edmonds–Karp algorithm for computing maximal flows. *J. Ass. Comput. Mach.* **19**, 184–92.

—— (1973a). More pathological examples for network flow problems. *Math. Program.* **5**, 217–24.

—— (1973*b*). A bad network problem for the simplex method and other minimum cost flow algorithms. *Math. Program.* **5**, 255–66.

ZYKOV, A. A. (1949). On some properties of linear complexes. *Math. Sb.* **24**, 163–88. English translation: *Am. Math. Soc. Translation No. 79*, 1952.

Subject index

abbreviation, 88
absorption law, 26
absorptive graph, 96–7
absorptive matrix, 103–5, 129, 137, 244
accessible set, 43
  labelling algorithm for, 45–7
activity graph, 51–2, 110, 150
acyclic graph, 48–53, 98–9, 109, 149–50
adjacency, 33
adjacency matrix, 97–105
  Boolean, 98, 99, 100, 102, 104, 149–50
  closure of, 102–5, 121–3
  of acyclic graph, 98–9, 149–50
  powers of, 99–101
alphabet, 14
anti-reflexive graph, 36
anti-symmetric graph, 36
arborescence, 53
arc, 32
  basic, 146–7, 173
  entry, 226
  exit, 226
  inverted, 202
  normal, 202
  return, 206
  saturated, 208
articulation node, 165–71
articulation set, 165
ascendant, 43
assignment problem, 186–7, 213–14, 234
  optimal, 223–4, 234
associative law, 11
augmenting-chain algorithm, 184–7, 197
auxiliary network, 226–9

back-substitution, 109, 111, 126
backtrack programming method, 64–75, 78, 83
  for colouring a graph, 192–5, 197
  for cut sets of edges, 174
  for elementary cycles, 83

for elementary paths, 65–70, 78–9, 83, 141
for feedback node sets, 82, 83, 242–3
for Hamiltonian cycles, 71–5, 83
for maximal cliques, 177–81, 197
for spanning trees, 163–4, 174
basis, of a language, 88
basis graph, 148–50, 173–4
Bellman's method, for shortest paths, 125
bijective function, 9
bilateral connection, 235, 249
binary tree, 56
bipartite graph, 186–7
block,
  of a graph, 167–70, 174
  of a program, 152
block graph, 169
Boolean algebra, 30–1, 85, 93, 140–1
branch-and-bound methods, 83, 194, 197, 243
bridge, 156–8, 173–4

cancellation property, 15, 90, 137, 244
capacity,
  of a cut, 210–11
  of a node, 235, 249
  of an arc, 201
cardinality, of a set, 1
Cartesian product, 5
chain, 42–3
  alternating, 184
  augmenting, 184
  closed, 42
  elementary, 42
  open, 42
  simple, 42
chord, 146
  detection in acyclic graphs, 150
chromatic index, 195, 197
chromatic number, 80, 188–9, 196, 197
circuit, 42
  Hamiltonian, 71, 80, 83
circuit-edge connectedness, 156

Author index